QUANTUM CELLULAR AUTOMATA

Theory, Experimentation and Prospects

T0325146

QUANTUM CELLULAR AUTOMATA

Theory, Experimentation and Prospects

Editor

Massimo Macucci

Università di Pisa, Italy

Imperial College Press

Published by

Imperial College Press
57 Shelton Street
Covent Garden
London WC2H 9HE

Distributed by

World Scientific Publishing Co. Pte. Ltd.
5 Toh Tuck Link, Singapore 596224
USA office: 27 Warren Street, Suite 401-402, Hackensack, NJ 07601
UK office: 57 Shelton Street, Covent Garden, London WC2H 9HE

British Library Cataloguing-in-Publication Data
A catalogue record for this book is available from the British Library.

QUANTUM CELLULAR AUTOMATA
Theory, Experimentation and Prospects

ISBN-13 78-1-86094-632-5
ISBN-10 1-86094-632-1

Printed in Singapore

PREFACE

This book presents the results achieved in the investigation of the Quantum Cellular Automata (QCA) concept, a novel approach to the physical implementation of logic computation, which proposes a departure from the three-terminal device paradigm that has been at the center of automated computing since its inception. Most of the activities that will be discussed have been performed within the QUADRANT (QUAntum Devices foR Advanced Nano-electronic Technology) project, funded from 1997 to 2000 by the European Commission within MEL-ARI (Microelectronics Advanced Research Initiative), but also in the framework of a line of activity within another European project, ANSWERS (Autonomous Nanoelectronic Systems With Extended Replication and Signalling), and, as far as the work at Notre Dame University is concerned, within research programs funded by the National Science Foundation, the Semiconductor Research Corporation, the U.S. Defense Advanced Research Projects Agency, and the Office of Naval Research.

Partners in the QUADRANT project were the University of Pisa (Italy), the University of Cambridge (UK), the University of Linköping (Sweden), the CNRS-L2M laboratory (France), the University of Barcelona (Spain), the University of Tübingen (Germany), McMaster University (Canada), and the University of Notre Dame (USA).

The unifying theme of the book is defined by the approach followed throughout the QUADRANT activity: a continuous, two-way interaction between experimental and theoretical groups, giving rise to the creation of new knowledge and of new methodologies. In particular, we shall highlight efforts toward better integration of modeling, fabrication, and characterization activities, pointing out how such a synergy is instrumental in the development of a new technology.

Although interesting and innovative from a theoretical point of view, the QCA concept may be quite difficult to implement for large scale computation, at least in its initial formulation based on the Coulomb interaction between electrons. Even though the chances of application to a new generation of computers are not too big in the short term, research on QCA devices has significantly contributed to a better understanding of architectural and device requirement issues in the field of nanoelectronics. The book will convey to the reader these concepts in a structured fashion and in the form of a sequence of results derived from the QCA research, but valid in general for nanoelectronics.

The first chapter contains an introduction to the QCA concept as originally proposed, as well as a discussion on the extensions, in particular clocking, that have since appeared in the literature. The issue of power gain, fundamental for proper propagation of logic signals, is also addressed.

Basic aspects of modeling are presented in the second chapter, where the idealized model of a QCA cell and of coupled QCA cells is treated, on the basis of a Hubbard-like Hamiltonian, in which cells are described by means of a limited set of parameters. Such a Hamiltonian can be diagonalized, determining the ground state of the system, and therefore the cell polarization condition.

In order to treat realistic situations, with the inclusion of nonideality effects, more refined models such as those discussed in the third chapter are needed, where we start with the presentation of analytical techniques for the determination of the electrostatic potential and of the electron density in semiconductor nanostructures of interest for QCA implementation. We then move to the Configuration-Interaction technique, which allows proper representation of the bistable behavior of a cell, and discuss why methods that do not include the proper antisymmetrization of the wave function, such as the density functional approximation or the Hartree approximation, cannot reproduce cell bistability. In addition, we examine the operation of cells with more than two electrons and show that an occupancy of $4N + 2$ electrons, with N integer, allows proper QCA action.

Detailed quantum modeling cannot be extended to circuits consisting of several cells, because of the excessive resulting computational complexity. As in the mature field of microelectronics, we need to resort to a hierarchy of simulation tools, treating QCA circuits with semiclassical approaches which have been validated by comparing their outcome with that of refined quantum models for a few relevant cases. These semiclassical models, based on the minimization of the total electrostatic energy, are dealt with in the fourth chapter, where the time-independent simulation of circuits made of up to a few tens of QCA cells is treated in detail, along with analytical models that can be developed for chains of QCA cells of arbitrary length.

For practical applications it is also essential to be able to predict the time-dependent evolution of QCA cells, which is analyzed in detail in the fifth and sixth chapters, with two different models: a quantum model with the inclusion of dissipation phenomena in Chapter 5 and a semiclassical model based on a Monte Carlo technique in Chapter 6. The simulation approach developed in Chapter 6 has been applied to the determination of the maximum achievable speed of operation, and is suitable for treating generic single-electron circuits.

The experimental approaches to the implementation of QCA cells undertaken in QUADRANT are based on semiconductor technology, and are presented in Chapters 7 and 8, along with the simulation techniques that have been used for the design of the samples and the interpretation of measurement results. In particular, Chapter 7 is focused on Silicon-On-Insulator (SOI) technology, while Chapter 8 deals specifically with the GaAs/AlGaAs material system. In these chapters we provide also an overview of specific simulation techniques, with a comparison between their results and the experimental data.

An essential tool for the readout of the state of a QCA cell is represented by the noninvasive charge detector, that allows to measure the charge in a quantum dot with the resolution of a single electron, with a negligible perturbation. In Chapter 9 we discuss one such detector, based on the modulation of the conductance of a quantum point contact biased in the tunneling regime, which has been applied to the investigation of a QCA cell in GaAs technology.

So far, the most successful approach to the fabrication of actual QCA circuits has been the one based on metal tunnel junctions, which, although still limited to operating temperatures in the tens of millikelvins range, has allowed the demonstration of a binary wire, a majority voting gate, a shift register, and a few other basic configurations. Chapter 10 offers an overview of the work performed at Notre Dame University on this topic, offering evidence of power gain in clocked QCA circuits.

The more speculative and currently more promising ideas for the application of the QCA paradigm are based on molecular building blocks or on nanomagnetic systems. Molecular QCA systems, discussed in Chapter 11, offer an opportunity for ultimate downscaling, while the nanomagnetic implementation, presented in Chapter 12, should open new perspectives for overcoming the limitations imposed by electrostatic coupling in terms of insufficient energy splitting compared to the thermal energy.

Chapter 13 contains some concluding remarks and summarizes the lessons of general applicability that we have learned while carrying out research on the QCA concept.

Finally, I would like to acknowledge the continuous support received during the the execution of the QUADRANT and ANSWERS projects from the European Commission officers, Dr. Konstantinos Glinos and Dr. Ramón Compañó, who provided competent advice and enthusiastic encouragement.

M. Macucci
Università di Pisa
Pisa, Italy

CONTENTS

CHAPTER 1

The Concept of Quantum-Dot Cellular Automata

Craig S. Lent

Department of Electrical Engineering
University of Notre Dame
Notre Dame, IN 46556, U.S.A.

1.1. Needed: A New Device Paradigm for the Nanoscale

The first digital electronic computers were the result of two very good ideas: first, use binary numbers to represent information mathematically; and second, *physically* represent the binary "1" and "0" as the "on" and "off" states of a *current switch*. Konrad Zuse in the 1930s first used electromechanical relays as the current switches, and later changed to vacuum tube triodes. These were eventually replaced by the solid-state version, the semiconductor transistor. Modern CMOS involves a clever use of switches paired so that current flows only when the state of the pair changes. Representing binary information by turning current switches on or off has been one of the most fruitful ideas in the history of technology.

This current-switch paradigm does, however, have serious limitations as device sizes are reduced. As a switch shrinks, it becomes less able to turn the current off and on cleanly. Also, since the current through a single switch is small, it takes longer to charge the interconnect lines between devices. Charge quantization leads to large statistical current fluctuations. Finally, since electrons move from the power supply to ground, considerable energy dissipation occurs.

These limitations have come more fully into view as the shrinking of CMOS technology has continued its remarkable progress. Such fundamental considerations will ultimately limit the device densities attainable with transistors, although the precise trajectory of the microelectronics industry may be determined more by fabrication cost issues. One need not be committed to a particular forecast to see the growing importance of developing alternative approaches that would permit scaling electronics down to the ultimate limits of molecular dimensions.

The quantum-dot cellular automata (QCA) concept[1−17] involves keeping *one* of Zuse's ideas, using a binary representation of information, but replacing the current switch with a cell having a bistable charge configuration. One configuration of charge represents a binary "1," the other a "0," but no current flows into or out of the cell. The field from the charge configuration of one cell alters the charge configuration of the next cell. Remarkably, this basic device-device interaction, coupled with a clocking scheme for modulating the effective barrier between states, is sufficient to support general-purpose computing with very low power dissipation. QCA devices exist and multi-device circuits have been demonstrated at low temperatures. Work underway is aimed at developing a room-temperature molecular implementation of the QCA concept. The QCA approach, using molecules as structured charge containers, is a more natural match to molecular function than trying to use molecules as current switches.

1.2. The Physical Representation of Information

QCA represents a particular choice about how to represent information physically — use the charge configuration of a cell. It is helpful to compare this choice with alternatives, particularly at the molecular device level. Choices include representing information by:

1) Electronic charge state.
2) Electronic charge configuration (QCA).
3) Nuclear positions.
4) Electronic spin state.
5) Nuclear spin.
6) Collective magnetic moment.
7) Coherent electronic quantum state.
8) Superconducting ground state.

At the macroscopic level (1) can be viewed as the CMOS model. It is also possible to replace the gate of a transistor by a molecular charge center and encode information in its charge state. This has seen success in making memory applications.[18] Approach (3) has proven useful for making molecular memory by switching the conformational state of a molecule and sensing the state by resistance changes. This is effective for memory but inherently slow and limited in scaling by the size of the current contacts. Coherent quantum computing (QC) has used a combination of (5) and (7) in the gas phase and (4), (5), and (7) in some solid state proposals. The weakness is the well-known problem of decoherence. Since QC required isolated unitary time-evolution, but also measurement of the output, this problem appears fairly fundamental. Superconductor-based QC devices have combined (7) and (8), but are limited in temperature ($T < T_c$) and size (one must have enough carriers to form BCS ground-state condensate — not single molecules). Making necessarily generous assumptions of a long decoherence time, it has been shown that a coherent version of QCA is a computationally complete QCA system, but that does not solve the general problems of QC. Direct spin-spin coupling is too weak to support binary (non-QC) versions of QCA based on (4) or (5). Collective magnetic effects, while also limited in scaling, (6), are the basis of conventional memory systems, and a magnetic QCA logic approach is described elsewhere in this volume. While not small, magnetic systems[19] have the advantage of extremely high coupling energies. For molecular electronics, the combined features of speed, single-molecular size, and robust general-purpose computing argue for using approach (2).

1.3. Dots in QCA

The essential feature of a dot in a QCA cell is that it localizes charge. A dot is a region of space with potential barriers surrounding it which are sufficiently high or wide so that the charge within is, to a good approximation, quantized to a multiple of the elementary charge. Of course, at some point the barrier has to be transparent enough so that a charge can quantum mechanically tunnel from one dot to another. The intrinsic bistability of a QCA is a manifestation of the quantization of charge. It is important to note for various implementations of QCA the relationship between the single-particle quantized energy levels and the charging levels for the dots.

1.3.1. *Metal dots*

For the metal-dot QCA implementation the role of the dot is played by small metal islands on an insulating substrate. The single-particle energy levels are extremely close together in energy. Furthermore a single dot has billions of free electrons. But the Coulomb cost of one additional electron tunneling onto the dot ($\sim e^2/C$) may be quite large indeed. The charging of a dot is stabilized by this electrostatic effect and the single-particle spectrum is unimportant.

1.3.2. *Molecular dots*

Dots within a molecular QCA cell are simply redox centers within the molecule, areas of the molecule which can accept (be reduced) or donate (be oxidized) an electron without breaking the chemical bonds that hold the molecule together. Such dots have a very large single-particle energy level spacing and a high Coulomb cost for adding an additional charge. At the molecular level both of the two effects are strong.

1.3.3. *Semiconductor dots*

Semiconductor dots can be formed by electrostatically depleting a two-dimensional electron gas. Patterned metal on the surface is typically used to shape the confining potential. Alternatively, dots can be formed from self-assembled structures such as pyramids formed in appropriately tuned MBE growth. Dot sizes and separations are both of the order of tens of nanometers. At these length scales Coulomb energies and single-particle quantum confined energy levels are relatively small and comparable to each other. The single-particle energies can be separated by constricting the confining potential. The problem is that since the energy levels are very strongly affected by the size of the confinement ($E \sim 1/d^2$ for a square well), an extreme sensitivity to small variations in the geometry results. An early explored alternative was to form charge "puddles" in the semiconductor dots. If the dot contains a few tens of electrons, it behaves very much like a metal, with an addition energy dominated by the effective capacitance of the dot. It was seen, however, that in that case, using actual metals was more effective, since the capacitance between dots could be made larger and more controlled.

The Notre Dame group has focused on the metal and molecular implementations. In the metal-dot QCA cells, the energetics is dominated by the Coulomb interaction and well-described by a capacitance matrix. For the molecular QCA implementation a full quantum-chemistry calculation, including both the strong Coulomb interaction and many-electron quantum effects is required.

1.4. QCA Cells

The essential feature of a QCA cell is that it possesses an electric quadrupole which has two stable orientations. These two orientations are used to represent the two binary digits, "1" and "0." The simplest implementation of this is a 4-dot cell composed of dots at the corners of a square, with two mobile charges. The two quadrupole states then correspond simply to the two ways of occupying antipodal dots. The bit information is contained in the sign of the in-plane quadrupole moment. The polarization P of the cell is the normalized quadrupole moment. If there are no other fields, either from the environment or from the neighboring cells, then the two orientations of the quadrupole moment have the same electrostatic energy. The presence of nearby cells causes one arrangement to be the favored low-energy configuration.

It is clear that such a cell could also be considered as two "half-cell" dipoles which have opposite dipole moment. Indeed since the information is contained in a single such dipole, it can be considered the fundamental unit of a QCA cell. In most instances it is more convenient to consider the whole 4-dot cell as the basic unit. Using half-cells means keeping track of whether one has an even or odd number of (half) cells in order to correctly interpret the bit information.

Non-square rectangular cells are also possible. Square cells have the advantage of being easy to lay out in patterns that turn a right-angle corner without additional complication. In looking for self-assembled molecular QCA cells, one could relax this requirement.

1.5. The Quantum-Dot Cellular Automata Paradigm

An idealized QCA cell can be viewed as a set of four charge containers, or "dots," positioned at the corners of a square (Figs. 1.1(a) and 1.1(b)). The cell contains two extra mobile electrons which can quantum-mechanically

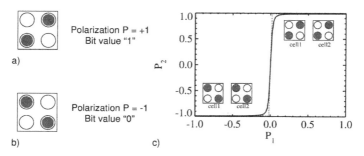

Fig. 1.1. QCA cells encode information in the charge configuration as shown in (a) and (b). The response of one cell to its neighbor is very nonlinear (c).

tunnel between dots but, by design, cannot tunnel between cells. The dots
can be realized in a number of ways: electrostatically formed quantum dots
in a semiconductor, small metallic islands connected by tunnel junctions, or
redox centers in a molecule. The barrier between dots should be high enough
so that charge can move only by tunneling and is therefore localized in the
dots and not in the connectors. The configuration of charge within the cell
is quantified by the cell polarization, which can vary between $P = -1$,
representing a binary "0," and $P = +1$, representing a binary "1," as
illustrated in the figure. The polarization of one cell induces a polarization
in a neighboring cell, purely through the Coulomb interaction. If the tunnel-
barriers are high, this interaction is very nonlinear (Fig. 1.1(c)).

Though the potential of the QCA concept extends beyond Boolean circuits, it is important that the approach be rich enough to support any
Boolean function. Circuits can be created by putting QCA cells in proximity to each other. A QCA *binary wire* is formed simply by creating a
linear array of cells as shown in Fig. 1.2(a). The Coulomb interaction makes
nearby cells align in the same state. The corner interaction is anti-voting
so it can be used to make an inverter (Fig. 1.2(b)). The natural logic *gate*
is the three-input majority gate shown in Figs. 1.2(c) and 1.2(d). We have
shown through extensive simulations that larger circuits can be designed
using hierarchical design rules.[2] A full adder has been simulated using the
full self-consistent Schrödinger equation, verifying that the adder works for
all input possibilities.

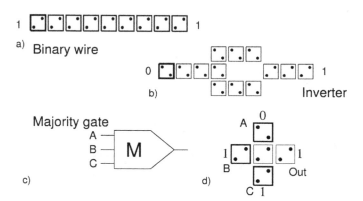

Fig. 1.2.   QCA devices.

1.6. Clocked QCA Cells

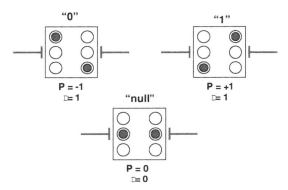

Fig. 1.3. Clocked QCA. The potential of the middle dots is varied by the clock potential, switching the cell from the null state to the active state.

A six-dot QCA cell as shown in Fig. 1.3 allows an additional level of control that has important implications for the way QCA cells can be used in circuits. A clocking signal varies the relative energies of the middle "null" dots and the active dots in the corners. If the clock signal makes the null dot energies much lower than those of the active dots, then the electrons are drawn into the null state regardless of the state of surrounding cells. If the clock signal raises the energy for electrons to be on the null dots, then they are forced out into the active dots. Whether they choose to be in the active "1" state or active "0" state depends on the state of the neighbors.

The state of the cell can be described in a simple way using two scalars. The polarization, P, and the activity α. If Q_{12} is the total charge on the null cells in units of the elementary charge, then the activity is simply $\alpha = 2 - Q_{12}$ for a six-dot cell. The activity can be varied from $\alpha = 0$ for a cell in the null state to $\alpha = 1$ for a cell in an active state with polarization $+1$ or -1.

The clocking signal can be implemented in a number of ways. For metal-dot cells it is simply a time-varying voltage applied to a lead which is capacitively coupled to the middle-dot islands. Because the metal-dot cells are physically large compared to molecular cells (and therefore cryogenic), it is possible to separately wire each cell to a clocking voltage. For molecular cells, it would be impossible to have a separate voltage connection to each molecule. Fortunately, another approach is possible. A local electric field perpendicular to the plane on which the molecules are arranged can provide the clocking signal.[3] This only requires that the middle null dots be at a different height above the plane than the active dots, yielding something like a "V" configuration.

1.7. Clocked QCA Shift Devices

A line of appropriately clocked QCA cells forms a shift register, a key component in QCA circuitry. We can consider first individually clocked cells for simplicity. As the clock signal is gradually raised, moving a cell from the null to active state, it polarizes into the same state as the cell to the left. This creates a copy of the bit shifted to the right. After the copy is made, the leftward cell can be erased to the null state, then brought back to the active state as it copies the succeeding bit in the bitstream. In this way information encoded in the cell polarization states is moved in a controlled way.[3] A continuously varying clocking wave can move bits of information smoothly along the line. In the molecular QCA case, the number of cells holding the bit information can be varied by varying the spacial wavelength of the clocking signal.

1.8. Power Gain

Power gain is essential for the practical operation of any information-processing system. As information moves from device to device and stage to stage, energy in the signal path is always lost to dissipative processes. These dissipative processes are unavoidable because the computing devices are always coupled to the rest of the environment. If there is not a way to replace the lost signal energy the signal will gradually degrade and ultimately be lost in the thermal noise. In conventional electronics the signal energy is automatically supplemented at each stage by energy from the power supply. In QCA circuits the energy is supplied automatically by the clock.

We can see the way gain comes about by considering a single cell in a QCA shift register which we will designate the target cell. Figure 1.4 illustrates the energy flow through a cell. In each clock cycle some energy, E_{in}, is transferred into the target cell from the neighbor to the left. That is, the neighboring cell does net work on the target cell. Similarly, the target cell transfers some energy E_{out} to the cell on its right. There is some loss to dissipative processes, E_{diss}, and some energy transferred from the clock to the target cell, E_{clock}. In steady state for a well-functioning shift register these are in balance so that $E_{out} = E_{in}$ and $E_{diss} = E_{clock}$ and the

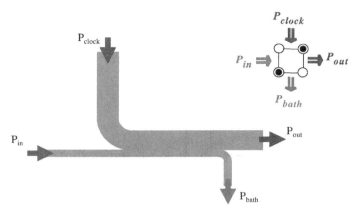

Fig. 1.4. Power flow in a QCA. Energy lost to dissipative processes is replaced by energy flowing in from the clock.

power gain is unity. If the input cell is weak, say because of a fabrication error or misalignment, then the input energy will be less than normal. In that case $E_{out} > E_{in}$ and $E_{diss} > E_{clock}$ so the clock is supplying additional energy to the signal path and the power gain is greater than one.[12] That this happens automatically, that is, as a consequence of the physical interactions themselves, has been verified by direct calculations and observed in experiments on metal-dot cells. Power gains greater than 3 have been measured.[13]

1.9. Robustness against Thermal Errors and Defects

The inherent bistability of QCA cells combines with the power gain possible with clocked QCA to yield considerable tolerance to both thermal errors and errors due to defects in patterning cell layouts. This can be seen clearly in the specific example of a shift register comprised of clocked metallic three-dot cells. Figure 1.5 shows the shift register and a unit cell. We consider cells with junction capacitors $C_j = 1.6$ aF, gate capacitors $C_g = 0.32$ aF, coupling capacitors $C_c = 0.8$ aF, and junction tunnel resistances $R_T = 100$ kΩ. The capacitance to ground is assumed to be the same as the gate capacitance. We consider a semi-infinite shift register with a four-phase clock. The equations of motion for the system are solved using a master equation approach with the transition rates given by the usual orthodox theory of Coulomb blockade devices. This allows us to examine the system response to (1) finite temperatures, (2) high speed operation,

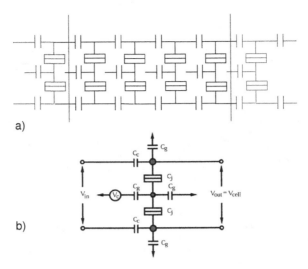

Fig. 1.5.   A metal-dot QCA shift register. (a) Schematic of shift register. (b) Unit cell of the shift register.

and (3) random variations in the values of all capacitances. The master equation approach lets us handle rare events precisely, an advantage over Monte Carlo techniques. A Monte Carlo calculation for the same system yields essentially identical results.

For low but non-zero temperatures (less than 5 K for the parameters chosen) a signal is transported an arbitrary distance down the shift register with no errors. The signal strength is defined as the potential between the top and bottom dot in the three-dot cell. As the temperature increases, the signal starts to decay as it moves down the line (Fig. 1.6) due to thermal fluctuations. As a function of temperature, the system exhibits behavior much like a phase transition. Below a critical temperature, there are no errors at all. Above that temperature errors inevitably accumulate. The abrupt nature of the transition can be seen by examining the deviation of the power gain from 1 (Fig. 1.7). At temperatures below the critical temperature, the power gain is precisely unity. Here the random thermal errors are always overcome by the power gain of the cell, restoring the system at each stage so no errors accumulate. Above the critical temperature, the power gain drops below unity rapidly as the temperature is raised.

A phase diagram can be constructed as a function of temperature and clock period (Fig. 1.8). An operating region which is entirely error-free is bounded by a critical temperature and a critical frequency (or equivalent

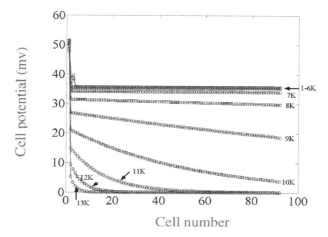

Fig. 1.6. Temperature dependence of metal-dot shift register. For temperatures below 5 K, the bit stream is transmitted to infinity with no errors. For higher temperatures, thermal excitations cause the signal to decay.

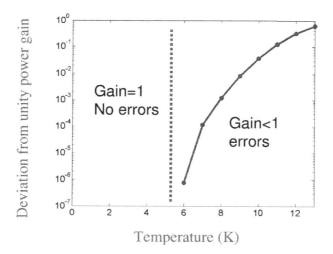

Fig. 1.7. Thermal dependence of power gain in shift register. Below a critical temperature the power gain is exactly one.

clock period). If the device is operated too fast or at too high a temperature, the power gain cannot keep up with the errors. But if it is operated below these critical values, signals can propagate to infinity with no thermal errors. If all the capacitors are scaled with the junction capacitance

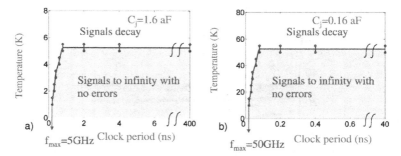

Fig. 1.8. Phase diagram for metal-dot QCA shift register. The operating region with no errors is bounded by a critical temperature and maximum frequency. The results scale with the capacitances.

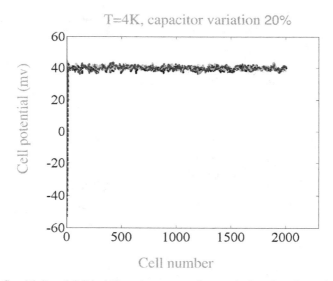

Fig. 1.9. Sensitivity of QCA shift register to random variations in values of capacitors.

to smaller values, the maximum temperature and maximum frequency increase accordingly.

If we now choose an operating point in the thermal error-free region, we can examine the effect of randomness in the capacitances. Each capacitor in a chain of 200 cells is varied randomly and the fidelity of signal transmission assessed. For variations of 20% (Fig. 1.9) bits are transmitted with no errors. If the variation is increased to 30% (Fig. 1.10), fidelity is

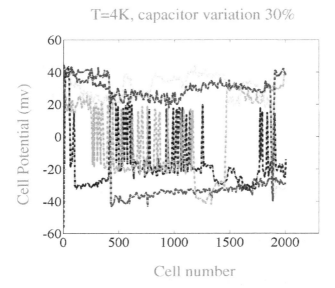

Fig. 1.10. Sensitivity of QCA shift register to random variations in values of capacitors.

lost. For molecular cells, the variations in capacitances would correspond to positional variations. Error tolerance can be additionally improved greatly by using wires which are more than one cell wide. Such triple-modular redundancy (or higher for wider wires) can overcome even gross errors and missing cells. Systematic study of such strategies is ongoing. Of course the metal-dot cells considered here remain low-temperature devices; achieving room-temperature operation requires scaling down sizes to molecular length scales.

1.10. Conclusions

QCA represents a new approach to digital devices which is still in its infancy. QCA devices do exist in the form of metal-dot QCA circuits. Though they require cryogenic operation, they form an important prototype system which may have niche applications. Molecular QCA will work at room temperature.

From the architectural point of view, the development of clocked QCA has been critically important. Clocking allows true power gain, which has now been demonstrated experimentally. It also allows us to begin to construct a computer architecture appropriate for QCA at the molecular scale. Clocking can be accomplished using straightforward fabrication of metallic electrodes (clocking wires) which serve to guide bit information around the circuit. This approach connects timing and layout in a more fundamental way than present CMOS architectures.

The power gain in clocking also permits the restoration of signal levels in a noisy thermal environment with fabrication errors. The combination of gain and redundancy (multi-cell wide wires) can yield considerable robustness against errors.

References

1. C. S. Lent, P. D. Tougaw, and W. Porod, Appl. Phys. Lett. **62**, 714 (1993); C. S. Lent, P. D. Tougaw, W. Porod, and G. H. Bernstein, Nanotechnology **4**, 49 (1993).
2. P. D. Tougaw and C. S. Lent, J. of Appl. Phys. **75**, 1818 (1994).
3. C. S. Lent and P. D. Tougaw, Proc. of the IEEE **85**, 541 (1997).
4. K. Hennessy and C. S. Lent, J. Vac. Sci. Tech. B **19**, 1752 (2001).
5. A. O. Orlov, I. Amlani, G. H. Bernstein, C. S. Lent, and G. L. Snider, Science **277**, 928 (1997).
6. C. S. Lent and P. D. Tougaw, J. of Appl. Phys. **75**, 4077 (1994).
7. I. Amlani, A. Orlov, G. Toth, G. H. Bernstein, C. S. Lent, G. L. Snider, Science **284**, 289 (1999).
8. A. O. Orlov, I. Amlani, C. S. Lent, G. H. Bernstein, and G. L. Snider, Appl. Phys. Lett. **74**, 2875 (1999).
9. A. O. Orlov, I. Amlani, R. Kummamuru, R. Rajagopal, G. Toth, C. S. Lent, G. H. Bernstein, and G. L. Snider, Appl. Phys. Lett. **77**, 295 (2000).
10. A. O. Orlov, R. Kummamuru, R. Ramasubramaniam, G. Toth, C. S. Lent, G. H. Bernstein, G. L. Snider, Appl. Phys. Lett. **78**, 1625 (2001).
11. G. Toth and C. S. Lent, J. Appl. Phys. **85**, 2977 (1999).
12. J. Timler and C. S. Lent, J. Appl. Phys. **91**. 823 (2002).
13. A. O. Orlov, I. Amlani, R. Kummamuru, R. Rajagopal, G. Toth, J. Timler, C. S. Lent, G. H. Bernstein, and G. L. Snider, Appl. Phys. Lett. **81** 1332 (2002).
14. M. T. Niemier and P. M. Kogge, International Journal of Circuit Theory and Applications, **29**, 49 (2001); M. T. Niemier and P. M. Kogge, International Conference on Electronics, Circuits, and Systems (ICECS '99), Cyprus, September 1999.
15. I. Amlani, A. O. Orlov, G. L. Snider, and G. H. Bernstein, Journal of Vacuum Science and Technology B **15**, 2832 (1997).
16. G. L. Snider, A. O. Orlov, I. Amlani, G. H. Bernstein, C. S. Lent, J. L. Merz, and W. Porod, Semi. Sci. & Tech. **13**, A130 (1998).
17. R. K. Kummamuru, A. O. Orlov, R. Ramasubramaniam, C. S. Lent, G. H. Bernstein, and G. L. Snider, IEEE Transactions on Electron Devices **50**, 1906 (2003).
18. Qiliang Li, Guru Mathur, Mais Homsi, Shyam Surthi, Veena Misra Vladimir Malinovskii, Karl-Heinz Schweikart, Lianhe Yu, Jonathan S. Lind-

sey Zhiming Liu, Rajeev B. Dabke, Amir Yasseri, David F. Bocian, Werner G. Kuhr, Appl. Phys. Lett. **81**, 1494 (2002).

19. R. P. Cowburn and M. E. Welland, Science **287**, 1466 (2000).

CHAPTER 2

QCA Simulation with the Occupation-Number Hamiltonian

Massimo Macucci and Michele Governale[a]

Dipartimento di Ingegneria dell'Informazione
Università di Pisa
Via Caruso 16, I-56122 Pisa, Italy

2.1. Introduction

Most of the work which exists in the literature on multiple quantum dot systems is based on extremely simplified models exploiting an occupation number (Hubbard-like) formalism. The detailed electronic structure of the single quantum dots is neglected and each dot and its interaction with the other dots are described by means of a few phenomenological parameters, such as the tunneling energy, the dot confinement energy, the on-site electrostatic interaction. Such descriptions are successful in capturing the overall behavior of the system, and in providing a qualitative understanding of the underlying physics, but do not allow accurate quantitative predictions and the development of effective design tools. Tougaw *et al.* have studied[1] the cell-to-cell response function by means of an occupation-number Hamiltonian and have derived results for a series of idealized cases. Conduction through coupled quantum dots has been studied by several authors within the framework of a Hubbard-like formulation: Stafford and Das Sarma have studied the appearance of collective Coulomb Blockade phenomena in arrays of quantum dots,[3] Kotlyar and Das Sarma have investigated transport through coupled double quantum dots[4] and persistent currents in quantum dot arrays,[5] while Klimeck *et al.* have studied the characteristics of conductance peaks in double-dot systems.[6]

---

[a]Current address: Institut für Theoretische Physik III, Ruhr-Universität Bochum, D-44780 Bochum, Germany

The approach based on the occupation-number Hamiltonian is rather straightforward to implement and requires very limited computational resources, however it has the drawback of involving the introduction of a phenomenological parameter, the tunneling energy t, which has to be determined with approximate methods.

Our model QCA is made up of four quantum dots placed at the corners of a square. In this initial approach we have limited the study to circular dots and to a total of just two electrons in the cell. In addition, we have assumed that only one state is allowed in each dot, and that tunneling can occur only between nearest neighbor dots. Gallium Arsenide material parameters have been considered, with an effective mass $m^* = 0.067 m_0$ and a relative permittivity $\varepsilon_r = 12.9$.

Our aim is that of computing the ground state charge distribution in a cell, subject to the electrostatic interaction of a nearby cell with a given charge distribution. Since the electrons in a cell tend to occupy dots belonging to the same diagonal axis, it is convenient to define the cell polarization, a quantity indicating the extent to which charge has accumulated along a preferential axis. Following Ref. [1], we define the polarization P as

$$P = \frac{(Q_1 + Q_3) - (Q_2 + Q_4)}{Q_1 + Q_2 + Q_3 + Q_4}, \tag{1}$$

where Q_i is the total charge in the i−th dot (in the following chapters we will introduce a different expression, which is more suitable to describe the electrostatic action of a cell on its neighbor).

2.2. Formulation of the Occupation-Number Hamiltonian

The Hamiltonian for a single, isolated cell reads

$$H_0 = \sum_{i,\sigma} E_{0,i} n_{i,\sigma} + \sum_{i>j,\sigma} t(b_{i,\sigma}^\dagger b_{j,\sigma} + b_{j,\sigma}^\dagger b_{i,\sigma}) \tag{2}$$
$$+ \sum_i E_{Qi} n_{i,\uparrow} n_{i,\downarrow} + \sum_{i>j,\sigma,\sigma'} V_Q \frac{n_{i,\sigma} n_{j,\sigma'}}{|\vec{R}_i - \vec{R}_j|},$$

where $E_{0,i}$ is the ground state energy of the single, isolated i−th dot, $b_{j,\sigma}^\dagger$ and $b_{j,\sigma}$ are, respectively, the creation and annihilation operators for an electron in the j−th dot with spin σ, $n_{j,\sigma}$ is the number operator for electrons in the i−th dot with spin σ, t is the tunneling energy between adjacent dots, V_Q is equal to $e^2/(4\pi\epsilon)$, with e being the electron charge and ϵ the dielectric permittivity of the medium, E_{Qi} is the on-site charging energy for the i−th dot,[3] $\vec{R}_i$ is the position of the center of the i−th dot. The tunneling energy t can be related to the actual potential landscape confining the dots only by means of some approximation. One typical approach is that of computing t as the semi-difference of the level-splitting deriving from lowering the barrier between the dots from infinity down to the actual value.

When a cell is coupled to a driver cell, another term needs to be added to the Hamiltonian, representing the electrostatic interaction from the driver cell:

$$H_{\text{int}} = \sum_{i \in \text{cell1}} \sum_{j \in \text{cell2}} V_Q \frac{\rho_{j,2} - \bar{\rho}}{|\vec{R}_{j,2} - \vec{R}_{i,1}|}, \tag{3}$$

where $\rho_{j,2}$ represents the average number of electrons in the j−th dot of cell 2 (the driving cell), $\bar{\rho}$ is the average number of donors per dot, $\vec{R}_{j,2}$ and $\vec{R}_{i,1}$ are the positions of the j−th dot of cell 2 and the i−th dot of cell 1, respectively.

2.3. Diagonalization of the Occupation-Number Hamiltonian

The total Hamiltonian is diagonalized using an occupation number representation: $|n_{1,\uparrow}, n_{1,\downarrow}; n_{2,\uparrow}, n_{2,\downarrow}; n_{3,\uparrow}, n_{3,\downarrow}; n_{4,\uparrow}, n_{4,\downarrow}\rangle$.

If we consider only the two-electron states, we obtain a vector space of dimension 256, but not all these states are important for our purposes. Indeed, when looking for the ground state, we can only consider the cases of electrons with opposite spin, and thus only 16 states are possible:

$$
\begin{aligned}
&|1,0;0,1;0,0;0,0\rangle \\
&|0,1;1,0;0,0;0,0\rangle \\
&|1,0;0,0;0,1;0,0\rangle \\
&|0,1;0,0;1,0;0,0\rangle \\
&\quad \cdots \\
&|1,1;0,0;0,0;0,0\rangle \\
&|0,0;1,1;0,0;0,0\rangle \\
&|0,0;0,0;1,1;0,0\rangle \\
&|0,0;0,0;0,0;1,1\rangle .
\end{aligned} \tag{4}
$$

These states are orthogonal and normalized to unity, thus they form a valid basis for our problem. Rewriting the Hamiltonian in the representation just described, we obtain a symmetric sparse matrix with only 4 nonzero off-diagonal elements in each row, whose eigenvalues and eigenvectors can be obtained with a standard numerical procedure. The ground state of the driven cell can then be expressed as

$$|\psi_0\rangle = \sum_{i=1}^{16} \alpha_i |i\rangle, \tag{5}$$

where α_i is the $i-$th element of the eigenvector corresponding to the lowest eigenvalue and $|i\rangle$ is the $i-$th element of the base. The average number of electrons in each dot is given by

$$\rho_i = \langle \psi_0 | n_{i\uparrow} + n_{i\downarrow} | \psi_0 \rangle, \tag{6}$$

from which we can then obtain the polarization of cell 1.

2.4. Application to the Evaluation of the Effects of Geometric Asymmetry on the Cell-to-Cell Response Function

We report results for the cell-to-cell response function (i.e. the polarization of the driven cell versus that of the driving cell) in the presence of an asymmetry in the driven cell. Results for the basic case of a perfectly symmetric cell, have been extensively discussed in the literature.[1,2] In Fig. 2.1 the polarization P_1 of the driven cell is plotted versus the polarization P_2 of the driver cell, for two cells separated by a distance d_c of 32 nm (this is the distance between the centers of the dots of cell 1 nearest to those of cell 2) and fabricated with gallium arsenide. The tunneling energy between pairs of nearby dots belonging to the same cell is 0.1×10^{-3} eV and the separation between the centers of such dots is 24 nm. The diameter D of the dots is 16 nm, except for one dot, whose diameter varies between 15.94 and 16.06 nm, and represents the parameter classifying the various curves. Even a very small asymmetry leads to a breakdown of the cell-to-cell response, preventing the correct operation of the QCA chain. This extreme sensitivity to cell symmetry is in contrast with more optimistic preliminary results that we had obtained on the basis of a comparison between the electrostatic interaction energy and the variation in the confinement energy associated with the asymmetry. The reason for this apparent contrast lies in the fact that we should instead compare the variation in the confinement energy and the energy separation between the two polarization configurations, which is very small. In Fig. 2.1 the effect of the asymmetry appears to

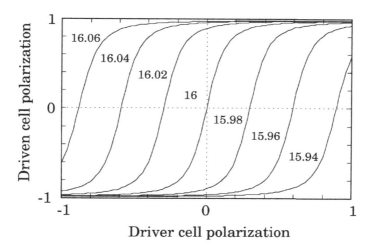

Fig. 2.1. Polarization of the driven cell P_1 versus the polarization of the driver cell P_2 as a function of the driver cell asymmetry.

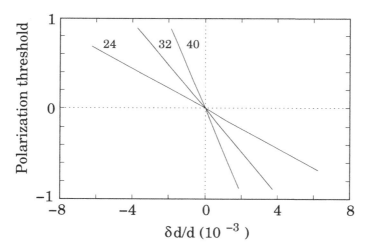

Fig. 2.2. Shift of the zero of polarization P_1 (threshold for switching) as a function of the driver cell asymmetry.

be mainly that of shifting the polarization curves along the horizontal axis. Therefore, the value for which the polarization of the driven cell changes sign moves from its zero value corresponding to a perfectly symmetric cell. We have defined the amount of this shift P_{sw} and we have reported it as a

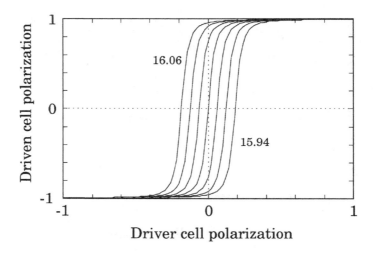

Fig. 2.3.   Polarization of the driven cell P_1 versus the polarization of the driver cell P_2, as a function of the driver cell asymmetry for a SOI QCA.

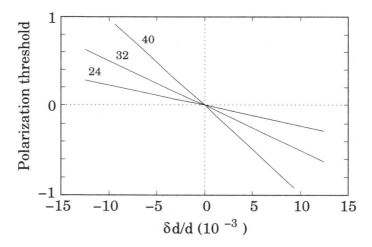

Fig. 2.4.   Shift of the zero of polarization P_1 (threshold for switching) as a function of the driver cell asymmetry for a SOI QCA.

function of the percentage dot diameter variation in Fig. 2.2. The various curves are for different values of the separation d_c between cells: the closer the cells the less disruptive is the effect of the asymmetry. It is possible to show that the results of Fig. 2.2 are independent of the tunneling energy.

In practice the situation is expected to be a little better, because the smooth shape of the confinement potential reduces the sensitivity to lithographic errors. A realistic model for a QCA cell and its implications for the effects of asymmetries are presented in detail in Chapter 3. For the silicon-on-insulator material system the situation is quantitatively different, since the effective mass in silicon dots is much larger than that in gallium arsenide and the relative dielectric permittivity of silicon oxide is around 4, about a third of that for gallium arsenide. An exact solution for the SOI material system would require a three dimensional model, because the electrostatic interaction has to be evaluated in the presence of a medium with varying permittivity (silicon and silicon oxide have very different permittivities): we have performed an approximate calculation using an average relative permittivity of 8 (which is actually an overestimate, since the electric field lines are mainly in the oxide region). Due to the larger effective mass, the confinement energy is smaller, while the electrostatic interaction is larger, as a consequence of the reduced permittivity. This leads to less sensitivity to the relative variation in the dot size, as shown in Fig. 2.3 and in Fig. 2.4, where the results for the SOI system are shown, for the same choice of geometric parameters as for the GaAs material system, for which results have been presented in the two previous figures. Thus it is expected that requirements on fabrication tolerances will be less strict for SOI cells than for cells based of GaAs heterostructures. This advantage, however, has to be weighed against increased fabrication difficulties for SOI, as will be discussed in detail in the following chapters.

References

1. P. D. Tougaw, C. S. Lent, and W. Porod, *J. Appl. Phys.* **74**, 3558 (1993).
2. P. Douglas Tougaw and Craig S. Lent, *J. Appl. Phys.* **75**, 1818 (1994).
3. C. A. Stafford, S. Das Sarma, *Phys. Rev. Lett.* **72**, 3590 (1994).
4. R. Kotlyar and S. Das Sarma, *Phys. Rev. B* **56**, 13235 (1997).
5. R. Kotlyar and S. Das Sarma, *Phys. Rev. B* **55**, R10205, (1997).
6. G. Klimeck, G. Chen, and S. Datta, *Phys. Rev. B* **50**, 2316 (1994).
7. J.H. Davies, *Semicond. Sci. Technol.* **3**, 995 (1988).
8. R. MacWeeny, *Methods of molecular quantum mechanics*, London, Academic Press (1989).
9. J.C. Slater, *Quantum theory of Matter*, McGraw-Hill, New York (1968).
10. M. Macucci, A. Galick, and U. Ravaioli, *Phys. Rev. B* **52**, 5210 (1995).
11. M. Field, C. G. Smith, M. Pepper, D. A. Ritchie, J. E. F. Frost, G. A. C. Jones, and D. G. Hasko, *Phys. Rev. Lett.* **70**, 1311 (1993).

CHAPTER 3

Realistic Time-Independent Models of a QCA Cell

Joan Martorell

Dept. d'Estructura i Constituents de la Materia
Facultat de Física,
Universitat de Barcelona
E-08028 Barcelona, Spain

Donald W. L. Sprung

Department of Physics and Astronomy
McMaster University
L8S 4M1 Hamilton, Ontario, Canada

Michele Girlanda[a] and Massimo Macucci

Dipartimento di Ingegneria dell'Informazione
Università di Pisa
Via Caruso 16, I-56122 Pisa, Italy

3.1. Introduction

We present several approaches to the time-independent realistic simulation of QCA cells, with models that are capable of yielding quantitative agreement with experiments. We start with the discussion of analytic approaches which clarify the physics involved in the more complex numerical simulations, and are also useful as an independent crosscheck for the results given by the codes. Pioneering work along this line is due to John Davies and collaborators in Glasgow.[1,2] We expand on some of that work,

[a]Current address: Dipartimento di Chimica e Chimica Industriale, Università di Pisa, Via Risorgimento, 35, I-56122 Pisa, Italy

and also on our results published in Refs. 3 and 4 and further developed in the course of QUADRANT.[5] We focus on methods to determine the depletion under a stripe gate. The complementary problem of a wire induced by a split gate has been recently reviewed by Raichev and Debray.[6] We then move on to the description of the Configuration-Interaction approach that we have used for the numerical simulation of a complete cell, and present its application to the analysis of the effects of fabrication tolerances. We introduce a cell layout that allows correction of the geometrical asymmetries by properly tuning the bias voltages applied to the gates and discuss how such a layout can be useful for the demonstration of the QCA principle, but cannot be applied to the realization of large-scale circuits. We also apply the Configuration-Interaction method to the analysis of cells with more than 2 electrons, and show that correct operation is possible with $4N + 2$ electrons, where N is an integer. Finally, we study a chain of cells defined with the previously introduced layout, and demonstrate that correct propagation of polarization can be achieved.

3.2. Heterostructure with a Uniform Gate

We start with the simplest configuration: a delta doped heterostructure composed of cap, donor, spacer and substrate layers, with a metallic gate covering the entire surface. The y-axis is chosen perpendicular to the surface, with the origin $y = 0$ on the surface. From the Poisson Thomas-Fermi results of Ref. 3, Eq. (27), we find that then the equilibrium electron density, n_e, is determined by the voltage applied to the gate, V_g, via the relation:

$$eV_g = -eV_s + \frac{e^2}{\epsilon}(n_d y_i - n_e y_s) - \left(\frac{e^2}{\epsilon}\kappa\, n_e\right)^{4/5}$$

$$\text{with} \quad \kappa^2 \equiv \frac{15\pi^2\epsilon}{4e^2}\left(\frac{\hbar^2}{2m^*}\right)^{3/2}, \tag{1}$$

where y_i denotes the location of the delta doped layer and y_s the position of the spacer-substrate junction. In the model it is assumed that the Schottky barrier height is eV_s, so that the semiconductor conduction band edge at the surface is at $eV_s + eV_g$ above the Fermi level. The ionized donor sheet density is n_d. It is a fixed fraction of the total donor density, whose amount will depend on how the device has been prepared. Assuming that the experiment is performed at low temperature T, the electrons trapped in DX centers cannot be ionized by changes in V_g.

In particular, setting $n_e = 0$ in Eq. (1) we find the pinch off potential

$$eV_{p.o.} = -eV_s + \frac{e^2}{\varepsilon} n_d y_i \ , \tag{2}$$

whereas the electron density at zero voltage is determined from the same equation, solving:

$$0 = -eV_s + \frac{e^2}{\epsilon}(n_d y_i - n_{e,0} y_s) - \left(\frac{e^2}{\epsilon}\kappa n_{e,0}\right)^{4/5} \ . \tag{3}$$

This allows rewriting the pinch off potential as:

$$eV_{p.o.} = \frac{e^2}{\epsilon} n_{e,0}\left(y_s + \frac{\kappa^{4/5}}{(e^2/\epsilon n_{e,0})^{1/5}}\right) \equiv \frac{e^2}{\epsilon} n_{e,0} y_e \ . \tag{4}$$

The last term in the parentheses acts as an effective distance between a purely two dimensional electron gas and the junction.

However, we shall be more interested in the linear response regime, around the equilibrium electron density at zero gate voltage, $n_{e,0}$. For sufficiently weak applied gate voltage, by linearizing Eq. (1) we find:

$$\delta n_e \equiv n_e - n_{e,0} = -\frac{eV_g}{y_e^{(l)} e^2/\epsilon} \ , \tag{5}$$

where we have introduced a different effective 2DEG (two-dimensional electron gas) location:

$$y_e^{(l)} \equiv y_s + \frac{4}{5}\frac{\kappa^{4/5}}{(e^2/\epsilon \, n_{e,0})^{1/5}} \ . \tag{6}$$

These well known results serve as an introduction to the methods that will be applied in the following to more complex configurations.

3.3. Linear Gate

We will consider the same heterostructure as above. The linear gate parallel to the y-axis has width $2a$. The origin is on the surface, on the symmetry axis of the gate. Two widely used models for the exposed surface are Fermi level pinning and "frozen surface".

We start with **Fermi level pinning:** This model assumes that the surface states can exchange charge with the 2DEG so that both have the same Fermi level when there is no potential applied to the gate. The corresponding equilibrium electron density is denoted $n_{e,0}$, as above. When a negative potential $V_g < 0$ is applied to the gate it is assumed that, 1) the ungated portion of the surface remains at zero voltage (as would be the case for a perfect insulator), 2) the depletion of the 2DEG is determined by the condition that it behaves as a perfect classical metal, so the potential is the same at all points, pinned by the states of the exposed surface. The electrostatic potential due to V_g applied to the gate must therefore satisfy the following boundary conditions:

$$V_{el}(x,0) = V_g \quad \text{when} \quad |x| < a \quad , \quad = 0 \quad \text{when} \quad |x| > a$$
$$V_{el}(x,d) = 0 \, , \tag{7}$$

where d is the location of the effective two-dimensional electron gas. Solving the Laplace equation with these boundary conditions is a standard problem in conformal mapping, and the result is given in Ref. 2. An alternative expression, which makes the dependence on the various parameters more explicit, can be derived as follows:

With the above choice of origin, the linear gate corresponds to the interval $(-a, a)$ on the x-axis, and the 2DEG is located on the plane $y = d$. For convenience we introduce dimensionless variables: $X = \pi x/d$, $Y = \pi y/d$ and $\Phi = V_{el}/V_g$. In the new variables, the gate is located in the interval $(-A, A)$, $A = \pi a/d$, and the 2DEG is in the plane $Y = \pi$.

Let $z = X + iY$, and consider the conformal mapping: $w = \exp(z)$. This takes the X-axis into the positive part of the $u = \text{Re}(w)$ axis, whereas the line $Y = \pi$ gets mapped into the negative part of the u axis. The strip between the $Y = 0$ and $Y = \pi$ lines is mapped onto the upper half w plane.

In the new variables, the boundary conditions on $\Phi(u, v)$ are:

$$\Phi(u,0) = 1 \qquad \text{when} \qquad u_l \equiv e^{-A} < u < u_r \equiv e^{A}$$
$$\Phi(u,0) = 0 \qquad \text{otherwise} \, . \tag{8}$$

An analytic function fulfilling these boundary conditions has the form

$$\Phi = \text{Im} \left\{ \alpha \ln(w - u_l) + \beta \ln(w - u_r) + \gamma \right\} \, . \tag{9}$$

We write $w - u_l = M e^{i\theta_l}$ in polar form: $w - u_r = N e^{i\theta_r}$. Imposing Eq. (8) one finds: $\alpha = -\beta = 1/\pi$, $\gamma = 0$. To write Φ in terms of X and Y we note that:

$$\theta_r = \arctan \frac{v}{u - u_r}$$
$$\theta_l = \arctan \frac{v}{u - u_l}$$
$$u = e^X \cos Y$$
$$v = e^X \sin Y \tag{10}$$

and therefore:

$$V_{el}(x,y) = V_g \Phi(X,Y) = \frac{V_g}{\pi} \left(\arctan \left(\frac{v}{u-u_r} \right) - \arctan \left(\frac{v}{u-u_l} \right) \right)$$

$$= \frac{V_g}{\pi} \left(\arctan \left(\frac{e^X \sin Y}{e^X \cos Y - e^A} \right) - \arctan \left(\frac{e^X \sin Y}{e^X \cos Y - e^{-A}} \right) \right) . \quad (11)$$

This is the desired explicit expression for the electrostatic potential. To determine the depleted 2DEG charge, we require that when $y > d$ the total electric field should vanish. Then the electric field due to the change in electron density, $\delta n_e(x) = n_e(x) - n_{e,0}$, must compensate that due to the applied potential:

$$\frac{e \, \delta n_e}{\epsilon} - \frac{\partial V_{el}}{\partial y} \bigg|_{y=d+0} = 0 . \quad (12)$$

Differentiating Eq. (11) with respect to y:

$$\pi \left[\frac{\partial \Phi}{\partial Y} \right]_{Y=\pi} = -\frac{\sinh A}{\cosh A + \cosh X} = \quad (13)$$

$$- \frac{\tanh(A/2)}{1 + \sinh^2(X/2)/\cosh^2(A/2)} ,$$

so that:

$$\delta n_e(x) = -\frac{\epsilon V_g}{ed} \frac{\tanh(\frac{1}{2}\pi a/d)}{1 + \mathrm{sech}^2(\frac{1}{2}\pi a/d) \sinh^2(\frac{1}{2}\pi x/d)} , \quad (14)$$

which expresses $\delta n_e(x)$ as a function of V_g. In the limit $a \to \infty$ the result for the uniform gate, Eq. (5), is recovered, provided that one chooses $d = y_e^{(l)}$ with the latter given by Eq. (6).

In Fig. 3.1 we compare a set of numerical two dimensional electron densities, $n_e(x)$, extracted from 3D Poisson Thomas-Fermi calculations, to the predictions of the model just described. For the latter we use $d = y_e^{(l)}$ as defined by Eq. (6). The simulations were made for a device grown at Bagneux by Jin *et al.*, with a 2DEG at the depth of 35 nm and a measured density of $n_{e,0} = 0.0051 \; nm^{-2}$. The ionized charge n_d was adjusted so that Poisson Thomas-Fermi calculations did reproduce this value. Although we have not used any additional adjustable parameter in the model, the agreement with fully 3D PTF is excellent. This supports the validity of the assumptions introduced in the model.

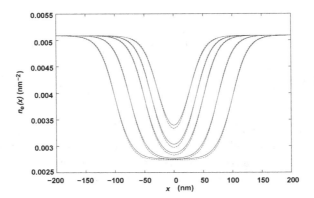

Fig. 3.1.   Two dimensional electron density under a linear gate at $V_g = -0.14$V and gate widths: $w = 50, 75, 100, 150$ and 200 nm. Fermi level pinning predictions: the continuous line corresponds to a 3D Poisson Thomas-Fermi calculation, while the dashed line corresponds to the linear approximation of Eq. (14).

The above analytical result, Eq. (14), can be simplified by noting that for most cases of interest one can consider $\pi a/(2d) >> 1$, and therefore:

$$\frac{\tanh(\tfrac{1}{2}\pi a/d)}{1 + \mathrm{sech}^2(\tfrac{1}{2}\pi a/d)\sinh^2(\tfrac{1}{2}\pi x/d)} \simeq \frac{1}{1 + e^{\pi(x-a)/d}} \ . \qquad (15)$$

This shows that the surface thickness of the depleted gas is governed solely by d, whereas the effective width of the depleted region is that of the gate: a.

A word of caution on the validity of the linear approximation: the pinch off value for V_g corresponds to $n_e(0) = 0$, and thus to:

$$eV_{p.o.} = \frac{e^2}{\epsilon}n_{e,0}d\frac{1}{\tanh(\pi a/(2d))} \ . \qquad (16)$$

For large a the hyperbolic tangent goes to 1. To recover the correct value, Eq. (4) one has to make a different choice for the location of the two dimensional gas, $d = y_e$.

Frozen surface approximation: In this case we assume that the normal component of the electric field vanishes on the exposed surface, so that the boundary conditions are:

$$V_{el}(x, y = 0) = V_g \quad \text{when} \quad |x| < a \quad \text{and} \quad \left.\frac{\partial V_{el}}{\partial y}\right|_{x,y=0} = 0 \quad \text{when} \quad |x| > a$$
$$(17)$$

and as before:

$$V_{el}(x, y = d) = 0 .\tag{18}$$

The most convenient dimensionless variables now are different:

$$X = \frac{\pi}{2}\frac{x}{d} \quad , \quad Y = \frac{\pi}{2}\frac{y}{d} \quad , \quad A = \frac{\pi}{2}\frac{a}{d} \quad , \quad \Phi(x, y) = \frac{V_{el}(x, y)}{V_g} ,\tag{19}$$

and the desired solution for Φ is obtained through two successive conformal mappings: The first is

$$\zeta = \frac{\tanh Z}{\tanh A} \quad , \quad Z = X + iY .\tag{20}$$

This maps the strip $0 < Im Z < \pi/2$ into the upper half plane, putting the boundary conditions on $\Phi(\zeta)$ on the new real axis with *i*) $\Phi = 0$ when $-\infty < \zeta_x < -1/k$, and when $1/k < \zeta_x < \infty$ ($k \equiv \tanh A$); *ii*) $\partial\Phi = 0$ when $-1/k < \zeta_x < -1$ and $1 < \zeta_x < 1/k$; *iii*) $\Phi = 1$ when $-1 < \zeta_x < 1$. The second conformal mapping is:

$$w = \int_0^\zeta \frac{dt}{\sqrt{(1 - t^2)(1 - k^2 t^2)}}\tag{21}$$

where $k = \tanh A$ has to be in the interval $(0, 1)$. Under these conditions it can be shown[10] that $w(\zeta)$ maps the upper half plane into a rectangle whose vertices, $-K + iK', -K, K, K + iK'$, correspond to the points $-1/k, -1, 1, 1/k$. The origin gets mapped into the origin and the points at $\pm\infty$ map into the point where the rectangle crosses the imaginary axis. The two constants are

$$K = \int_0^1 \frac{dt}{\sqrt{(1 - t^2)(1 - k^2 t^2)}}$$

$$K' = \int_0^1 \frac{dt}{\sqrt{(1 - t^2)(1 - (1 - k^2)t^2)}} .\tag{22}$$

After this mapping the boundary conditions simplify to: $\Phi = 0$ on the upper side of the rectangle, $\partial\Phi = 0$ on the left and right sides, and $\Phi = 1$ on the lower side. One sees immediately that

$$f(w) = i - \frac{w}{K'} \quad \text{with} \quad \Phi(w) = Im\{f(w)\}\tag{23}$$

fulfills the boundary conditions and is therefore the desired solution. Combining the mappings,

$$\Phi = Im\left\{i - \frac{w}{K'}\right\} = 1 - \frac{1}{K'}Im\left\{\int_0^{\tanh Z/\tanh A} \frac{1}{\sqrt{(1 - t^2)(1 - k^2 t^2)}}\right\},\tag{24}$$

and

$$\frac{\partial \Phi}{\partial Y} = -\frac{1}{K'} Im \left\{ \frac{i(\cosh Z)^{-2}}{\tanh A} \left[\left(1 - \left(\frac{\tanh Z}{\tanh A} \right)^2 \right) (1 - \tanh^2 Z) \right]^{-1/2} \right\}. \tag{25}$$

We are interested in the value at $Z = X + i\pi/2$, so that:

$$\tanh Z = \frac{1}{\tanh X} \quad , \quad \cosh Z = i \sinh X \quad \text{and}$$

$$\frac{\partial \Phi}{\partial Y}\Big|_{y=d} = \frac{1}{K'} \left[1 + \frac{\sinh^2 X}{\cosh^2 A} \right]^{-1/2} = \frac{1}{K'} \sqrt{\coth A} \sqrt{\frac{\sinh 2A}{\cosh 2A + \cosh 2X}}$$

$$= \frac{\sqrt{\coth A}}{K'} \sqrt{FF_s(2X, 2A)} \tag{26}$$

which is Eq. (7.1) of Davies et al.[2]

In the last line we connect the shape to the symmetrized Fermi function of arguments $2X, 2A$.[7] In this regard, $FF_s(r/d, R/d)$ is very similar to the usual Fermi function of half-radius R and surface thickness d, except that it has zero slope at the origin. In view of Eq. (19), FF_s appears with arguments $2X = \pi x/d, 2A = \pi a/d$, so that the shape is very close to that of the Fermi level pinning approximation (Eq. (15)).

Using again Eq. (12), we now find:

$$\delta n_e(x) = -\frac{\epsilon V_g}{ed} \frac{\pi}{2K'} \left[1 + \frac{\sinh^2(\pi x/(2d))}{\cosh^2(\pi a/(2d))} \right]^{-1/2} , \tag{27}$$

and it is easy to check that the limit for $a \to \infty$ is the same as that of the previous case. In Fig. 3.2, we show the depletions predicted with this frozen surface model, assuming the same electron density as before for the ungated 2DEG and $d = y_e^{(l)}$. For comparison, the dashed line corresponds to the Fermi level pinning prediction with $w = 200$ nm (already shown in Figure 1.) One can see that the two different assumptions for the exposed surface conditions affect mainly the effective width of the depleted region.

In the Thomas-Fermi approximation for a two dimensional electron gas, we can relate the electron density, $n_e(x)$, to an effective depleting potential, $U(x)$, noting that:

$$E_F^{(2D)} = \frac{\hbar^2}{2m^*} k_F^2(x) + U(x) \tag{28}$$

and, since in 2D $n_e(x) = k_F^2(x)/(2\pi)$, we can write:

$$U(x) = E_F^{(2D)} - \frac{\pi \hbar^2}{2m^*} n_e(x). \tag{29}$$

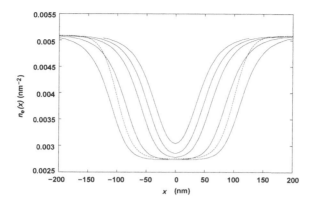

Fig. 3.2. Two-dimensional electron density under a linear gate at $V_g = -0.14$ V and for the gate widths: $w = 50, 75, 100, 150$ and 200 nm. The frozen surface predictions are shown with a continuous line, while a dashed line represents the Fermi level pinning prediction for $w = 200$ nm.

Therefore, the plots of $n_e(x)$ in Figs. 1 and 2 can be easily transformed into plots of the effective depleting potentials given by the Fermi level pinning and the frozen surface approximations, respectively. As can be seen, for the same V_g and gate parameters, the effective potentials have similar maxima, but differ somewhat in their surface shapes.

3.4. Linear Gate Deposited on Etched Surface

Some devices grown at Bagneux by Jin *et al.* had the gates deposited on an etched surface. This alters the boundary conditions. In this section we study the effect this will have on the conduction band edge inside the semiconductor. We will assume Fermi level pinning.

We begin by ignoring the contributions to the electrostatic potential of the ionized donors and the 2DEG; they will be incorporated later. We use conformal mapping techniques to determine solutions of the Laplace equation in the space below the etched surface, with boundary conditions fixed on it. Let us denote the plane of the unetched surface as $y_e = 0$, and place the origin of coordinates on that plane, so that the gate is located symmetrically with respect to it. The etched portion of the surface extends from $(x_e = -W/2, y_e = c_e)$ to $(x_e = W/2, y_e = c_e)$, and the metallic gate is deposited in the trench. Here c_e is the etching depth ($\simeq 2$ to 5 nm in the experiment). To define the mapping it is convenient to introduce two complex planes, one where the complex variable is $z_e = x_e + iy_e$, which is mapped onto a second plane, the z plane, where we write $z = x + iy$. The profile of the etched surface is defined by the following points (which are the vertices of an open Schwarz Christoffel (SC) polygon that will be used for the conformal mapping): $w_0 = -\infty + i0$, $w_1 = -W/2 + i0$, $w_2 = -W/2 + ic_e$, $w_3 = +W/2 + ic_e$, $w_4 = +W/2 + i0$ and $w_5 = +\infty + i0$. We are going to map the part of the upper half plane above this polygon into the upper half of the z plane. We map the sides of the polygon onto the x axis, choosing the images of the w_i as follows: $y_i = 0$ and $x_0 = -\infty$, $x_1 = -a$, $x_2 = -1$, $x_3 = 1$, $x_4 = a$, $x_5 = \infty$.

Following the usual rules for SC conformal mappings, one finds:

$$z_e = w(z) = A \int_0^z \sqrt{\frac{z^2 - 1}{z^2 - a^2}} \, dz + ic_e \tag{30}$$

where:

$$A = \frac{W/2}{\int_0^1 \sqrt{\frac{1-x^2}{a^2-x^2}} \, dx} \tag{31}$$

and the value of a is determined from:

$$\frac{W}{2} \int_1^a \sqrt{\frac{x^2 - 1}{a^2 - x^2}} \, dx = c_e \int_0^1 \sqrt{\frac{1 - x^2}{a^2 - x^2}} \, dx \ . \tag{32}$$

Solving numerically for one of the cases of interest: $c_e = 2.5$ nm, $W = 80$ nm, one finds $a = 1.03692$, $A = 43.49504$.

With this transformation, imposing Dirichlet b.c.'s in the z plane:

$$eV_L(x, y = 0) = 0 \quad \text{when} \quad |x| > a$$
$$eV_L(x, y = 0) = eV_g \quad \text{when} \quad |x| < a \ , \tag{33}$$

one can construct the potential in the mapped plane:

$$eV_L(x, y) = \frac{eV_g}{\pi} \left(\arctan \frac{x + a}{y} - \arctan \frac{x - a}{y} \right) \tag{34}$$

and therefore the potential in the physical plane is:

$$eV(x_e, y_e) = eV(x(x_e, y_e), y(x_e, y_e)) \ . \tag{35}$$

The functions $x = x(x_e, y_e)$ and $y = y(x_e, y_e)$ must be obtained from the mapping in Eq. (30).

To add the electrostatic potentials due to the 2DEG and the ionized donor layer we proceed in a manner analogous to that introduced in Ref. 8 for linear wires, but here we have to take into account the distortion introduced by the conformal mapping.

First note a property of any conformal mapping is that for any harmonic function Φ:

$$\frac{\partial^2 \Phi}{\partial x^2} + \frac{\partial^2 \Phi}{\partial y^2} = |w'(z)|^2 \left(\frac{\partial^2 \Phi}{\partial x_e^2} + \frac{\partial^2 \Phi}{\partial y_e^2} \right). \tag{36}$$

In particular, when $\Phi = -2q \ln |w - w_0|$, with $|w - w_0| = \sqrt{(x_e - x_{e,0})^2 + (y_e - y_{e,0})^2}$, one has:

$$\frac{\partial^2 \Phi}{\partial x^2} + \frac{\partial^2 \Phi}{\partial y^2} = -|w'(z)|^2 q \delta(w - w_0) = -q\delta(z - z_0), \tag{37}$$

where in the last step we use $\delta(w(z) - w_0) = |w'(z_0)|^{-2}\delta(z - z_0)$. This proves that the appropriate $\Phi(x, y)$ is

$$\Phi(x, y) = -q \ln((x - x_0)^2 + (y - y_0)^2) \tag{38}$$

and that therefore the combination of the Hartree and the mirror potential on an electron at the position (x_e, y_e) is given by

$$\begin{aligned} & e\, V_e(x(x_e, y_e), y(x_e, y_e)) \\ & = -\frac{e^2}{\varepsilon_r} \int_{-\infty}^{\infty} dy' \int_{-\infty}^{\infty} dx' \, \tilde{\rho}_e(x', y') \ln \frac{(x - x')^2 + (y - y')^2}{(x - x')^2 + (y + y')^2}, \end{aligned} \tag{39}$$

where $\tilde{\rho}_e$ is related to the electron density $\rho_e(x_e, y_e)$ by the charge conservation condition:

$$N_e = \int \tilde{\rho}_e(x, y)\, dx\, dy = \int \rho_e(x_e, y_e)\, dx_e dy_e = \int \rho_e(x_e, y_e) |w'(z)|^2 dx\, dy \tag{40}$$

so that

$$\tilde{\rho}_e(x, y) = \rho_e(x_e, y_e) |w'(z)|^2. \tag{41}$$

Combining these results:

$$\begin{aligned} & e\, V_e(x_e, y_e) = e V_e(x(x_e, y_e), y(x_e, y_e)) \\ & = -\frac{e^2}{\varepsilon_r} \int_{-\infty}^{\infty} dy'_e \int_{-\infty}^{\infty} dx'_e \, \rho_e(x'_e, y'_e) \ln \frac{(x - x')^2 + (y - y')^2}{(x - x')^2 + (y + y')^2}. \end{aligned} \tag{42}$$

Similar expressions also apply to the ionized donors contribution and its corresponding mirror terms, V_i. The total electrostatic potential is then: $V_{el} = V_L + V_e + V_i$.

In Fig. 3.3 we show results from 3D Poisson Thomas-Fermi calculations for the same device parameters considered earlier. What is different now is

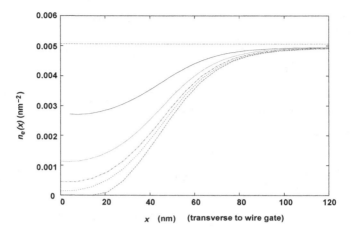

Fig. 3.3. Two dimensional electron density under a linear gate deposited on a surface etched to a depth $c_e = 2.5\ nm$. Gate width $W = 80$ nm and $V_g = 0, -0.1, -0.15, -0.17$, and -0.195 V.

that the gate has been deposited after etching the surface to a depth of $c_e = 2.5$ nm. The horizontal dashed line in the figure is the reference value for the 2DEG density for the ungated device. Due to etching, already at zero gate potential there is already almost 50% depletion at $x_e = 0$. This is shown by the first dotted line in the figure. The further depleted densities correspond to the applied gate potentials $V_g = -0.10, -0.15, -0.17$ and $-0.195\ V$. Note that for a shallow 2DEG the effect of etching is very important and cannot be ignored in the simulations.

3.5. Modeling of a Complete QCA Cell

The previously described approximations for the evaluation of the electrostatic potential produced by an arrangement of depletion gates can be combined with a "one-shot" technique for the solution of the Schrödinger equation, in order to obtain a detailed description of the cell response as a function of external perturbations.

We need a "one-shot" method because commonly used iterative procedures do fail when applied to a QCA cell: the reason for such a failure lies in the presence of a strong electrostatic interaction (which cannot therefore be considered a perturbation of the bare confinement potential, as is instead the case for small quantum dots) and of degeneracies connected with the fourfold symmetry of the structure. Initial attempts with Density Functional based approaches and with the Hartree and Hartree-Fock methods have exhibited very strong instabilities, lack of convergence, shifting of the electron density back and forth between two opposite locations as the iterations progress.

This is the reason why we decided to switch to an approach based on the Configuration-Iteration technique (CI), which allows a non-iterative solution of the combined Schrödinger-Poisson problem. Such a technique is based on the representation of the many-electron wave function as an expansion of a basis of Slater determinants. If the basis were infinite, this would be equivalent to the exact diagonalization of the Hamiltonian, but any practical implementation must be limited to a finite number of determinants: the number of determinants sufficient to obtain a reliable solution is strongly dependent on the characteristics of the problem. It is obvious that the number of basis determinants actually needed depends on how well the elements of the basis fit the solution wave function. As a general practical rule, we can state that, if the common choice of using as a basis determinants constructed with the wave functions for the case of noninteracting electrons is made, the number of required determinants is very dependent on the amount of deformation of the wave functions introduced by the electron-electron interaction, and therefore on the strength of such an interaction. A word of caution is needed — we can apply the CI method (which is typical of quantum chemistry) without too many problems because we are operating, as will become clear later, within the framework of a Fermi level pinning approximation, and therefore we are using a Dirichlet boundary condition at the exposed surface of the semiconductor, as far as the Poisson equation is concerned. This allows us to treat the electrostatic interaction as an interaction between point charges (with the inclusion of the images deriving from the Dirichlet boundary condition at the surface), for which a simple closed-form expression exists. In a general case, in which the complete Poisson equation must be solved, we would need to numerically evaluate the Green's function of the Poisson equation between each pair of points in the simulation domain, which, as the size of the domain is increased, quickly becomes a prohibitive task.

3.6. The Configuration-Interaction Method

Let us introduce the CI method by considering a generic Hamiltonian including a two-particle interaction $g(\vec{r}_i, \vec{r}_j)$:

$$\hat{H} = \hat{H}_1 + \hat{H}_2 \,, \tag{43}$$

$$\hat{H}_1 = \sum_{i=1}^{N}(-\frac{\hbar^2}{2\,m}\,\nabla_i^2 + V(\vec{r}_i)) = \sum_{i=1}^{N} h(\vec{r}_i) \,, \tag{44}$$

$$\hat{H}_2 = \sum_{i<j} g(\vec{r}_i, \vec{r}_j) \,, \quad \text{with} \quad g(\vec{r}_i, \vec{r}_j) = g(\vec{r}_j, \vec{r}_i) \,, \tag{45}$$

where $\hbar$ is the reduced Planck constant and m is the effective mass of the electron (we consider the case of Gallium Arsenide, with $m = 0.067 m_0$, m_0 being the free electron mass).

We then consider a numerable complete basis $\{\varphi_i(\vec{q})\}$, where with q we indicate both the spatial and the spin coordinates, over which the single-particle wave function for our problem can be expanded. We define the elements of such a basis "spin-orbitals".

Starting from the spin-orbitals, we construct all possible independent Slater determinants

$$\Phi_k = \frac{1}{\sqrt{N!}} \begin{vmatrix} \varphi_{n_{1k}}(\vec{q}_1) & \varphi_{n_{2k}}(\vec{q}_1) & \cdots & \varphi_{n_{Nk}}(\vec{q}_1) \\ \varphi_{n_{1k}}(\vec{q}_2) & \varphi_{n_{2k}}(\vec{q}_2) & \cdots & \varphi_{n_{Nk}}(\vec{q}_2) \\ \cdots & \cdots & \cdots & \cdots \\ \varphi_{n_{1k}}(\vec{q}_N) & \varphi_{n_{2k}}(\vec{q}_N) & \cdots & \varphi_{n_{Nk}}(\vec{q}_N) \end{vmatrix} \,, \tag{46}$$

where n_{ik} indicates which spin-orbital appears in the i-th column of the k-th Slater determinant. The N-electron wave functions Ψ_i of the Hamiltonian can be written, as already mentioned, as a linear combination of the Φ_k orbitals:

$$\Psi_i = \sum_{k=1}^{\infty} c_{i\,k} \Phi_k \,. \tag{47}$$

Computing the eigenfunctions of the Hamiltonian is therefore equivalent to finding the c_{ik} coefficients. If we define the vector $\vec{c}_i$ as consisting of the c_{ik} coefficients for a given i, the diagonalization of the Hamiltonian can be recast into the solution of the algebraic eigenvalue problem

$$\mathcal{H}\vec{c}_i = E_i \, \vec{c}_i \,, \tag{48}$$

where the elements of $\mathcal{H}$ are given by

$$\mathcal{H}_{k\,k'} = \langle \Phi_k | \hat{H} | \Phi_{k'} \rangle = \int \Phi_k^*(\hat{H}_1 + \hat{H}_2)\Phi_{k'} \prod_{i=1}^{N} d\vec{q}_i \,, \tag{49}$$

with E_i being the i-th eigenvalue of $\mathcal{H}$ and $\int d\vec{q}_i$ standing for integration over the spatial coordinates and summation over the possible spin orientations.

In principle this would be an infinite-dimensional eigenvalue problem, for which it would be impossible to find a numerical solution: in practice we can consider a finite number of Slater determinants, including those which are expected to contribute most to the solution. From an operational point of view, we take into consideration a finite set of M spin-orbitals, and from them we construct all the possible determinants, which, for an N-electron system, will be

$$\mathcal{N}_{SD} = \binom{M}{N}.$$
(50)

Thus the dimension of the $\mathcal{H}$ matrix is $\mathcal{N}_{sd} \times \mathcal{N}_{sd}$, and, as long as this number is not too large, a numerical solution becomes possible.

The calculation of the matrix elements of the Hamiltonian can be performed according to well-established procedures. The diagonal elements are given by

$$\langle \Phi_k | \hat{H} | \Phi_k \rangle = \sum_i \langle \varphi_{n_{ik}} | h | \varphi_{n_{ik}} \rangle$$

$$+ \frac{1}{2} \sum_{ij} (\langle \varphi_{n_{ik}} \varphi_{n_{jk}} | g | \varphi_{n_{ik}} \varphi_{n_{jk}} \rangle - \langle \varphi_{n_{ik}} \varphi_{n_{jk}} | g | \varphi_{n_{jk}} \varphi_{n_{ik}} \rangle),$$
(51)

where in general

$$\langle \varphi_i \varphi_j | g | \varphi_l \varphi_m \rangle = \int dq_1 \, dq_2 \, \varphi_i^*(q_1) \varphi_j^*(q_2) g(\vec{r}_1, \vec{r}_2) \varphi_l(q_1) \varphi_m(q_2).$$
(52)

For the nondiagonal elements, proper selection rules exist, which allow us to tell in advance which elements are certainly zero (Slater's rules[12]). A nondiagonal element of the type $\langle \Phi_k | \hat{H} | \Phi_{k'} \rangle$ does not vanish only in two cases: if Φ_k and $\Phi_{k'}$ differ by one single spin-orbital or by two spin-orbitals. Thus we obtain, in the case of a single spin-orbital difference,

$$\langle \Phi_k | \hat{H} | \Phi_{k'} \rangle = \langle \varphi_{n_{ik}} | h | \varphi_{n_{ik'}} \rangle$$

$$+ \sum_{j \neq i} (\langle \varphi_{n_{ik}} \varphi_{n_{jk}} | g | \varphi_{n_{ik'}} \varphi_{n_{jk}} \rangle - \langle \varphi_{n_{ik}} \varphi_{n_{jk}} | g | \varphi_{n_{jk}} \varphi_{n_{ik'}} \rangle),$$
(53)

and, in the case of a two spin-orbital difference

$$\langle \Phi_k | \hat{H} | \Phi_{k'} \rangle = \langle \varphi_{n_{ik}} \varphi_{n_{jk}} | g | \varphi_{n_{ik'}} \varphi_{n_{jk'}} \rangle - \langle \varphi_{n_{ik}} \varphi_{n_{jk}} | g | \varphi_{n_{jk'}} \varphi_{n_{ik'}} \rangle.$$
(54)

Such expressions are valid if the spin-orbitals which are in common between the two determinants appear in the same columns. Should this not be the case, it is sufficient to apply a proper sequence of permutations to the the the columns of one of the determinants (keeping in mind that permutations of odd order involve a change of sign).

After solving for the $\vec{c}_i$'s, the electron density everywhere in the cell can be computed from the simple expression

$$\rho_i(\vec{r}_j) = N \sum_{s_j} \int |\Psi_i(\vec{r}_j, s_j, \vec{q}_1, \ldots, \vec{q}_N)|^2 \, d\vec{q}_1 \ldots d\vec{q}_N , \qquad (55)$$

where s_j is the spin orientation coordinate and the integration is performed over all the $\vec{q}_i$'s, except for $\vec{q}_j$.

Let us now focus on the specific case we are considering for the QCA cell. The complete Hamiltonian will read

$$\hat{H} = -\frac{\hbar^2}{2m} \nabla_1^2 - \frac{\hbar^2}{2m} \nabla_2^2 + V_{\text{con}}(\vec{r}_1) + V_{\text{con}}(\vec{r}_2) + V_{\text{driv}}(\vec{r}_1) + V_{\text{driv}}(\vec{r}_2)$$

$$+ \frac{1}{4\pi\epsilon} \frac{e^2}{|\vec{r}_1 - \vec{r}_2|} - \frac{1}{4\pi\epsilon} \frac{e^2}{\sqrt{|\vec{r}_1 - \vec{r}_2|^2 + (2z)^2}} - \frac{1}{4\pi\epsilon} \frac{e^2}{2z} , \qquad (56)$$

where V_{con} is the confinement potential computed as previously described, V_{driv} is the Coulomb potential due to the charge distribution in the neighboring driver cell, the last two terms include the effects of the image charges, and z is the distance of the 2DEG from the surface of the heterostructure. In Eq. (56) the image charge contributions have been included with the last two r.h.s. terms, in order to be consistent with the approach used to compute the confinement potential, and their expressions have been obtained according to the following consideration: we take half the electrostatic energy of the system made up of the electrons and their image counterparts, as the energy stored in the image space is purely fictitious. The last term, due to the interaction of each electron with its own image, yields a constant shift of the energy spectrum.

We start from the simulation of a basic cell containing 2 electrons, and we choose a set of n functions $\{\psi_i(\vec{r})\}$ of the spatial coordinates (corresponding in our case to the single-electron wave functions), which we define "orbitals," in order to distinguish them from the spin-orbitals $\{\varphi_i(\mathbf{q})\}$. The spin-orbitals are obtained by combining each orbital with one of the two possible spin orientations, therefore we have $M = 2n$ spin-orbitals. The total number of Slater determinants that can be built with 2 electrons and

$2n$ spin-orbitals is $n(2n - 1)$. For our calculations we consider only n^2 determinants, because we exclude those which have a total spin different from zero.[5]

The matrix $\mathcal{H}$ that appears in the algebraic eigenvalue problem has therefore dimensions $n^2 \times n^2$ and is full, since, being the determinants of size 2×2, no pair of them can differ for more than two spin-orbitals. Since we have chosen real functions for the orbitals (the wave functions obtained solving the single-electron Schrödinger equation in confined region are real), $\mathcal{H}$ is real and symmetric.

With our choice of orbitals, we find out that a total of 12 are sufficient to construct a set of determinants over which a faithful expansion of the 2-electron wave function can be performed. It should be pointed out that such a limited number of orbitals suffices because the electrostatic interaction between the electrons does not lead to significant deformation of the wave functions with respect to those for the noninteracting system. The situation would be very different, as we shall see in the following, if there were more than one electron per dot: in such a case, the very strong electrostatic interaction would require usage of a much larger basis.

As a general rule, we consider the number of determinants as sufficient if a further increase of their number does not lead to an appreciable difference in the solution.

3.6.1. *Cell defined with a hole-array gate*

Let us begin with the analysis of the behavior of a cell obtained by means of a depletion gate with a circular hole for each dot (see Fig. 3.4), considering 90 nm holes placed at the vertices of a square, with a distance between centers of 110 nm.

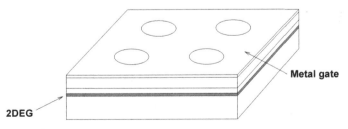

Fig. 3.4. Gate layout for the definition of a four-dot cell in the 2-dimensional gas obtained by modulation doping in a heterostructure.

We compute the cell-to-cell response function for a few values of the 2DEG depth and report the results in Fig. 3.5, for a gate bias voltage of −0.5 V. The driver cell is assumed to be fully polarized, located at a distance of 400 nm (distance between cell centers) from the cell being considered. It is apparent that the cell-to-cell response gets smoother as the depth of the 2DEG is increased: this does not come as a surprise, since, as the 2DEG gets further from the gate, the depth of the potential dips representing the dots decreases, thus the barriers seen by the electrons between the dots are lowered, and therefore a smoother transition of charge is expected to take place, as discussed in the previous chapter.

One of the main motivations in setting up this computational machinery has been the purpose of assessing the sensitivity of various implementations of QCA cells to fabrications tolerances. Therefore we have introduced, in the basic layout of the cell, an imperfection which is likely to occur: one of the holes in the gate has been made slightly smaller than the others, with a diameter corresponding to 0.9999 times the nominal value. This is a dimensional error orders of magnitude smaller than those which would be occurring in practice, and it is already sufficient to significantly perturb the cell-to-cell response function, as it is possible to see from Fig. 3.6, where the cell-to-cell response function is reported (solid lines) for three values

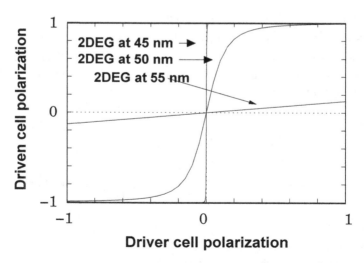

Fig. 3.5. Cell-to-cell response function for different values of the depth of the 2-dimensional electron gas.

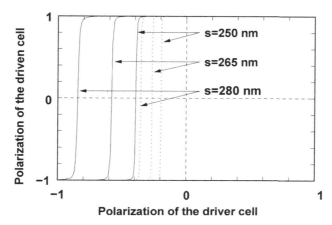

Fig. 3.6. Response function for a cell with 3 dots defined by 90 nm holes and one dot defined by a 89.991 nm hole. The holes are located at the corners of a square with a 110 nm separation between centers. Solid curves represent the response function for three values of the inter-cell distance (250, 265 and 280 nm). The dotted curves correspond to the response function for a situation in which the action of image charges has been neglected.

of the distance between cell centers: from 250 nm to 280 nm. The dotted lines represent, for comparison, the results obtained neglecting the effect of the image charges: the shift is less than in the case with images because the electrostatic interaction between cells is not screened in this case. It is apparent that, if we assume Fermi level pinning at the exposed surface, for a separation slightly larger than 280 nm the driven cell will be stuck in the configuration corresponding to a polarization $P = 1$, independent of the polarization of the driver cell. Even if we completely neglect the screening, a slightly larger asymmetry will still be sufficient to disrupt operation of the QCA cell.

If, instead of a variation in the diameter of one of the holes, we introduce an asymmetry associated with a displacement in the position of its center, our simulations show that a shift of 0.05 nm is already sufficient to prevent correct operation. The cell-to-cell response function for a 0.0275 nm shift is reported in Fig. 3.7. We point out that in a simplified model such as the one described in the previous chapter a shift in the position of one of the dots would yield a much smaller effect on the cell-to-cell response function, because it would not produce any deformation of the overall confinement potential: the only variation would be in the electrostatic interaction.

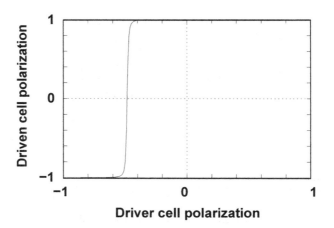

Fig. 3.7. Response function for a cell with the position of one of the dots shifted by 0.0275 nm with respect to the ideal geometry.

3.6.2. *Multiple-gate cell*

It is thus apparent that a functioning QCA circuit cannot be fabricated with an approach based on an array of holes defined in a depletion gate. In order to overcome such a problem, we have proposed[5] a different layout, based on multiple gates, whose bias voltages can be independently adjusted. The basic gate configuration is shown in Fig. 3.8: if properly biased, the seven electrodes define four quantum dots, approximately arranged at the corners of a square (or, more precisely, a rectangle). The presence of unavoidable fabrication errors would make the cell unusable if we applied the same voltages to all the gates (the situation would be analogous to that of the previously discussed hole array gate), however in this case we have several additional degrees of freedom: we can independently adjust the bias voltages, in such a way as to obtain an electrical asymmetry compensating for the geometrical asymmetry.

In order to verify whether such a correction is actually possible, we have developed an algorithm for the numerical evaluation of the set of gate voltages yielding proper symmetrization. We start from a symmetric choice of bias voltages, and then change one of the bias voltages at a time, monitoring the result on the basis of the split between the first and the fourth eigenvalues. Indeed, a perfect cell would be symmetric, and therefore would be characterized by a fourfold degeneracy: thus the first four eigenvalues would ideally be identical. This means that the difference in energy be-

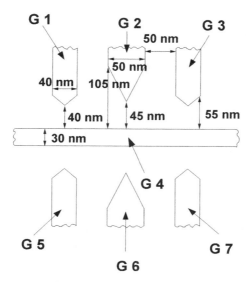

Fig. 3.8. Multi-gate layout for the definition of a four-dot cell with the possibility of electrical symmetrization.

tween the first and the fourth eigenvalue can be assumed as a measure of cell symmetry, and its decrease below a given threshold is a suitable stopping criterion for the iterative optimization procedure. We have been able to show that by properly adjusting the gate voltages,[5] correct operation can be obtained even for a 5 nm error in the position of one of the gates. In order to facilitate convergence of the numerical procedure, it is preferable to repeat the calculation several times, gradually increasing the position error: for example, in the calculations of Ref.[5] we have started with a displacement of 0.01 nm, which has then been increased up to 5 nm with a geometric progression.

By tuning the geometric asymmetry, it also possible to compensate for sources of nonideality other than geometric errors; for example, the potential perturbations due to the presence of randomly distributed dopants and of charged impurities trapped at the metal-semiconductor interface can be compensated, at least to the extent of restoring basic cell symmetry and functionality.

The layout of Fig. 3.8 can thus be used for an experimental demonstration of QCA action in a GaAs/AlGaAs heterostructure (and it has actually been used for this purpose [13]), but it also has significant limitations. Although cells can be coupled by aligning them along the horizontal direction, no lateral branching is possible, because the space above and below the cell is occupied by the leads needed to bias the individual gates. Therefore this layout cannot be used, at least in the present form, for actual QCA circuits, since lateral branching is required for the implementation of any type of logic.[14] Furthermore, the adjustment of the gate voltages required to symmetrize a cell can be performed on a circuit containing just a few cells, but is unthinkable for a large array of cells. In particular, we must consider that the tuning of one cell also depends on the bias voltages applied to the neighboring cells, as will be discussed more in detail at the end of the present chapter.

3.7. Analysis of Cells with more than 2 Electrons

In the early papers on the QCA concept,[15] operation of cells with more than 2 electrons was considered extremely inefficient or outright impossible: the main reason for this conclusion was use of the polarization expression

$$P = \frac{(\rho_1 + \rho_3) - (\rho_2 + \rho_4)}{\rho_1 + \rho_2 + \rho_3 + \rho_4}, \qquad (57)$$

where ρ_i represents the charge in the i-th dot (with respect to Ref. [15] we have neglected the term ρ_0 in the denominator, since we are discussing 4-dot cells instead of 5-dot ones). With such a polarization expression, it is apparent that increasing the number of electrons beyond 2 will certainly lead to a decrease of the maximum achievable polarization, since the difference at the numerator is at most equal to the charge of two electrons (basic electrostatics forbids configurations in which the charge along a diagonal is larger than that along the other diagonal by more than 2 electrons), while the denominator equals the charge of the total number of electrons in the cell.

This polarization formula, however, is not the best suited to express the action that a driver cell exerts on a driven cell. The global monopole component of the electric field produced by a cell is zero, because of the charge neutrality of a cell (the negative charge of the electrons is compensated by the positive charge of the ionized donors and by the charge induced on the metal gates), however such a neutralization occurs over a certain region, which leads to some effect dependent on the total number of electrons. Nevertheless, the prevalent action of a driver cell is the result of the multipole component resulting from the charge imbalance between the two diagonals: such a component varies between a minimum of zero and a maximum obtained when the imbalance equals twice the electron charge.

In other words, for a given imbalance between the diagonals, the action of a cell on its neighbor will be substantially the same for any value of the total number of electrons. Therefore the driving action is best expressed by the ratio of the imbalance to twice the electron charge (its largest possible value), as in the expression that we shall use in the following:

$$P = \frac{(\rho_1 + \rho_3) - (\rho_2 + \rho_4)}{2e}. \tag{58}$$

It is worth pointing out that if the positive neutralizing charge were located in the dots themselves, in an amount corresponding to $Ne/4$ (N being the total number of electrons in the cell), the electrostatic action of a cell would depend exclusively on the imbalance between the diagonals; in practice this is not exactly the case, because a significant contribution to the overall neutralization is given by charge outside the cell (induced on the gates and/or corresponding to ionized impurities), and some effect from the total number of electrons will therefore exist, although of limited importance in most cases.

Adoption of Eq. (58) is supported by the results of a calculation published in Ref. 16, where two limiting conditions were studied:

- i) the cell is neutralized with four Ne/q charges located at the same x, y coordinates as the dots, but on a plane at a vertical distance $z = d$ from that containing the cell;
- ii) the cell is neutralized with image charges located on a plane at a vertical distance $z = h$ from that of the cell (equivalent to enforcing a Dirichlet boundary condition at a distance $h/2$).

Condition (i) corresponds to neutralization by ionized donors that are located, as normally happens in a heterostructure with modulation doping, in a layer spatially separated from that of the quantum dots; condition (ii) corresponds instead to neutralization that is completely due to the charges induced on the metal gates and on the pinned surface of the semiconductor.

3.7.1. *Many-electron driver cell*

Let us first consider a driver cell with a variable number of electrons acting upon a driven cell with two electrons. We assume that the driven cell is defined, at a depth of 70 nm below the heterostructure surface, by a gate with four 90 nm holes located at the corners of a 100 nm square. The gate bias is -0.5 V and the distance between the driver and the driven cells is 300 nm.

If we assume neutralization according to condition (i), for a polarization 0.7 of the driver cell we obtain the polarization of the driven cell shown in Fig. 3.9 as a function of the distance d of the plane containing the positive charges. Three different curves are reported, for 2, 26 and 50 electrons in the driver cell: while with 2 electrons the polarization of the driven cell is full for any value of d, with more electrons the driven cell polarizes correctly only up to a certain distance of the plane containing the neutralizing charges. If such a plane is too far, the repulsive effect of the electrons in the driver cell is large enough to force the driven cell into an incorrect condition, i.e. a condition with both electrons in the two dots further away from the driver cell, which corresponds to no polarization. The considered electron numbers in the driver cell have been chosen corresponding to $4M + 2$, where M is an integer, because this is the only choice which allows, as will become apparent in the following, an imbalance of 2 electrons between the charges stored along the two diagonals.

In the case of neutralization with condition (ii), shown in Fig. 3.10, we have a more critical situation, because both for large and for small distances of the image plane the polarization of the driven cell is incorrect: for a large value of the distance failure is due to the same reason as in the previous case, while for small distances the driven cell does not polarize at all, because the dipole component from the driver cell is screened by the images. There is, however, a relatively wide interval of distance values for which correct operation is obtained. Notice that, although the curve for 50 electrons does

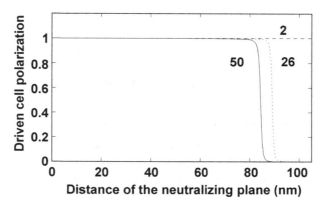

Fig. 3.9.   Polarization of a driven cell as a function of the distance from the image plane for a driver cell with an occupancy of 2, 26 and 50 electrons and a polarization of 0.7.

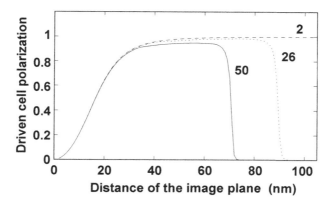

Fig. 3.10. Polarization of a driven cell as a function of the distance from the neutralizing plane, containing a charge $ne/4$ per dot, for a driver cell with an occupancy of 2, 26 and 50 electrons and a polarization of 0.7.

not reach the full polarization, we still have polarization gain, because this is the response to a driver cell with polarization 0.7.

Real situations are intermediate between the two limiting cases we have just presented, with neutralization partly due to donors and partly to the action of the gates and of the semiconductor surface. On the basis of these results, we can conclude that Eq. (58) is a proper representation of the action of the driver cell upon the driven cell for an arbitrary number of electrons.

3.7.2. *Semiclassical model*

In order to justify the choice we have made to consider situations with $4M + 2$ electrons, it is useful to consider a simple model based on classical electrostatics. Let us now analyze the behavior of a two electron driver cell coupled to a many-electron driven cell, and let us focus on the configurations of the driven cell corresponding to 3, 4, 5, and 6 electrons, as shown in Fig. 3.11. In a cell with 3 electrons, two of them will occupy the diagonal corresponding to the proper polarization, while the third will localize into the dot which (of the empty ones) is further from the driver cell. This corresponds, according to our expression, to a polarization of 0.5. If the driven cell contains 4 electrons, they will fill the 4 available dots, and polarization will vanish. If we add a fifth electron, this will end up in the dot furthest from the two occupied dots of the driver cell, leading to a polarization of

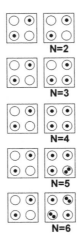

Fig. 3.11.   Configurations of a driven cell with 2, 3, 4, 5, 6 electrons, based on classical electrostatics.

-0.5. For 6 electrons, instead, the two dots along the diagonal corresponding to the correct polarization will be characterized by double occupancy, thus leading to a full polarization of 1. The same reasoning can be repeated for larger numbers of electrons, realizing that the same situation repeats upon addition of 4 electrons: from this simple model our conclusion is that correct operation, with full control of the driven cell, can be obtained when the number of electrons is $4M + 2$, with M integer.

3.7.3. *Many-electron driver cell*

We have performed a quantum calculation, based on the Configuration-Interaction method, to confirm such results: the cell layout is the same as for the previously presented investigation of a many-electron cell driving a two-electron cell, but with an applied gate voltage of -0.7 V instead of -0.5 V. In this case the driver cell has two electrons, while the driven cell has 2, 3, 4, 5, 6 electrons. We have used a larger negative gate bias because we need to compensate for the presence of an increased number of electrons, which leads to a rise of the self-consistent potential seen by each electron, thereby causing a reduction of the relative height of the inter-dot barriers. The results for the cell-to-cell response function are shown in Fig. 3.12: the limiting values of the driven cell polarization correspond to the predictions from the classical electrostatic model; we also notice that the response function for the 6-electron cell, although perfectly functional, is less steep than that of the 2-electron cell. This can be easily explained considering that the bare confinement potential is the same for both cases, and thus, as already mentioned, the self-consistent potential seen by the electrons of the 6-electron cell will be less deep (due to the contribution of the additional 4 electrons that give rise to a repulsive component), which leads to more transparent barriers and therefore to a less abrupt response function.

Hence, the possibility of operating a cell with a number of electrons larger than two is demonstrated, which is important from the experimental point of view, because it is extremely hard to achieve single-electron occupancy in quantum dots, while dots with a few tens of electrons can be routinely obtained.

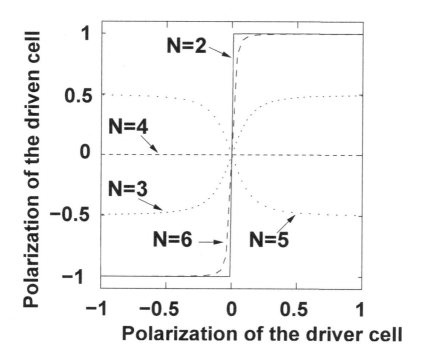

Fig. 3.12.   Cell-to-cell response function for a two-electron driver cell and a driven cell with 2, 3, 4, 5, 6 electrons.

3.8. Analysis of Polarization Propagation along a Semiconductor-Based Quantum Cellular Automaton Chain

Up to this point, we have presented results for the action of a driver cell (which is kept in a given polarization state with unspecified means) on a single driven cell. In the present section the actual techniques for enforcing the polarization of the driver cell will be discussed and a detailed numerical analysis of polarization propagation along a chain of three cells will be performed.

On the basis of Thomas-Fermi calculations,[23] it has been pointed out that the electrostatic action on the driven cells resulting from the polarization of the driver cell would be almost canceled by the very same gate bias imbalance applied to obtain such a polarization in the first place. This would result in a practical impossibility to obtain data propagation along a wire of QCA cells of any length.

However, although the Thomas-Fermi approach affords a proper description of the shape of the confinement potential, it does not take into account the discreteness of charge, which at the basis of the bistability of QCA cells:[14,5] this is the reason why a Thomas-Fermi approach leads to misleading conclusions if applied to the analysis of the bistable behavior and of polarization propagation.

In the following a realistic simulation of a three-cell QCA wire defined in a GaAs/AlGaAs heterostructure by means of lateral confinement with multiple metal gates will be presented, including a discussion of the choice of the proper bias voltage to be applied to the gates defining the driver cell in order to enforce the desired propagation of the polarization state.

3.8.1. Model of a three-cell chain

The gate layout for a linear chain (binary wire) is obtained by periodically repeating the gate structure of a single QCA cell, which has been presented earlier in this chapter.

A number of additional cells, besides the active ones, are included, in order to prevent symmetry breakdown arising from the sudden termination of the chain. The inclusion of the symmetrization cells is convenient in our numerical calculations, but in a practical implementation symmetrization could be achieved more easily by adjusting the bias voltages applied to a few additional gates placed in the proximity of the outer active cells.

A rather shallow heterostructure, with a 2DEG (two-dimensional electron gas) at a depth of only 35 nm, is considered, which allows formation of quantum dots of small dimensions. The heterostructure layer sequence is as follows: 10 nm GaAs cap layer, 15 nm AlGaAs layer, 10 nm undoped GaAs spacer layer and GaAs substrate. Within the hypothesis of Fermi level pinning at the semiconductor surface, we have chosen a pinning value of 0.65 eV below the conduction band, a surface charge density of 5×10^{12} electrons/cm^2 and a doping of 3.71×10^{18} cm^{-3} in the AlGaAs doped layer, which correspond to typical experimental parameters.

The layout defining a single cell has been repeated with a periodicity of 442 nm, with the inclusion, as shown in Fig. 3.13, of vertical stripes 152 nm wide kept at the same potential as the central electrode (D). Such stripes deplete the 2DEG between adjacent cells, thus preventing unwanted screening of the intercell interaction.

The electrodes defining a cell do perturb the confinement potential at the 2DEG level of a nearby cells, thus disrupting the symmetry unless a second analogous gate structure is placed on the opposite side. It has previously been shown that even a small perturbation of symmetry is sufficient to prevent operation of a cell, therefore in a line of three cells only the central cell would work properly (since the perturbing action of the two outer cels cancels out in the middle). In order to make them work, additional, "dummy" cells should be added on both sides, until each active cells "sees" the same gate structures in both directions.

Such a condition should be achievable in principle only including an infinite number of cells, because, as we add cells on one side, we create

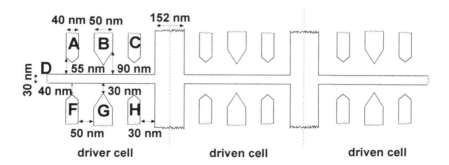

Fig. 3.13. Portion of the gate layout defining the three active cells (one driver and two driven). The voltage of the gates A, C, D, F. and H is set at -0.8 *V* while the one of the gates B and G is -0.9 *V*. The same gate layout replicates periodically to the left and to the right, defining a total of 21 cells.

an imbalance on the other side. However, since we are interested in the operation of a line of only three QCA cells and the electrostatic interaction drops as a function of distance, a finite number of dummy cells on each side of the chain will be sufficient for providing the required degree of symmetry for the potential seen by the three active cells located in the middle. After a few preliminary tests we have found that with a total of 21 cells we obtain a sufficient symmetrization of the potential.

We have defined the inactive cells used for balancing as "dummy," because they contain no electrons and do not contribute to polarization propagation. In an actual experiment, they could be made inactive by damaging the 2DEG in correspondence of the gates defining them or, as already stated, they could be replaced by a few, properly biased adjustment gates.

In Fig. 3.13, we report the gate layout only for the portion of the wire of interest (i.e. omitting the dummy cells). The extreme sensitivity to asymmetries would introduce very serious problems if, instead of a simple chain, a circuit with lateral branches were considered (once the previously mentioned layout limitations had been solved). In the presence of additional lateral branches, further electrodes should be added to achieve balancing, and their bias voltages should be adjusted in order to counterbalance the asymmetry introduced by the branch. It is apparent that the task of setting up a circuit of a certain complexity would be extremely challenging.

As already discussed for the case of a single QCA cell, the vertical confinement of the 2DEG due to the heterostructure can be assumed to be large enough to obtain occupation of only the lowest vertical subband, consequently a two-dimensional model can be used in our simulations. We have computed the confinement potential following the procedure discussed in Sec. 3.3 and we do not turn to a Schrödinger-Poisson calculation for the calculation of the exact occupancy, because the uncertainties in the heterostructure parameters (such as density of surface states, doping, presence of imperfections) would make the results unreliable anyway for our purposes. In a practical experiment the required occupancy would be obtained by means of careful tuning of the bias voltages.

In the following we deal with cells with double occupancy, in order to keep the calculation as simple as possible. In Fig. 3.13, the voltage bias for each gate in the first active cell is indicated. Gate voltage values are the same for all the cells taken into consideration, both dummy and active ones. This cell layout produces the confinement potential shown in Fig. 3.14 at a depth of 35 nm, with four well localized minima for each cell. The wide vertical branches of the D electrode (Fig. 3.13) assure that no 2DEG island

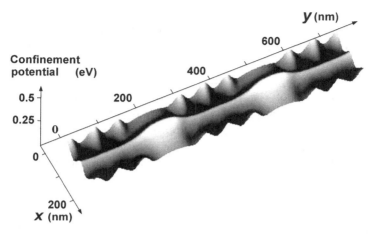

Fig. 3.14. Potential landscape defining the three active cells: the electrons are confined in the minima, marked by dark dots. Adapted with permission from M. Girlanda, M. Governale, M. Macucci, G. Iannaccone, *Appl. Phys. Lett.* **75**, 3198 (1999). Copyright 1999, American Institute of Physics.

is present between adjacent cells. In such conditions, since the charging energy must be at least ten times the thermal energy, operation is practicable only at a temperature below fractions of a kelvin. We have assumed zero temperature for all of our calculations, thereby neglecting any contribution from excited states.

We have computed the ground state following the procedure described in Sec. 3.6 based on the Configuration-interaction (CI) approach. Finally, we argue that coupling between different cells occurs only through the electrostatic interaction, since the electron wave functions vanish in the regions between different cells. Therefore the Schrödinger equation is solved with the CI approach within each four-dot cell.

The first active cell is the one whose polarization is enforced from the outside and that induces the polarization of the following two. For this reason it is indicated as "driver cell" while the remaining two cells are are denoted as (first and second) "driven cells."

The problem whether and how polarization of the driver cell can be enforced is not trivial and has been debated in the literature (see, e.g., Ref. 23). We have selected an approach consisting in polarizing the driver cell with the addition of a small positive term δV to the bias voltage of its lower right and upper left electrodes while a correspondent term $-\delta V$

is added to the bias voltage of its upper right and bottom left gates. This creates an imbalance in the confinement potential and forces the two electrons to localize in the bottom right and upper left dots. Obviously, it is sufficient to reverse the sign of δV to obtain the opposite polarization state.

The simulation of the three active cells is undertaken with a self-consistent procedure that starts with the computation of the electron density in the driver cell by means of a CI approach, neglecting the contribution of the electrons in the two other cells. The resulting electron density is used to calculate the potential acting on the first driven cell and then to perform a CI computation for such a cell (Fig. 3.15(a)).

The successive steps of the procedure are schematically illustrated in Figs. 3.15(b-d) where arrows indicate that the density arising from the electrons in one (or two) cell(s) is taken into account in the calculation of the ground state of the successive cell(s). The size of the dots in the driver cell indicates which dots are subject to a repulsive (smaller ones) force and which ones are instead attracting electrons (larger ones) as a consequence of the bias imbalance. Shading denotes occupied dots.

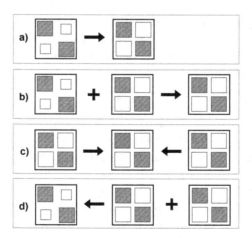

Fig. 3.15.   Schematic representation of the procedure used to reach self-consistency in the simulation of electrostatic coupling between cells: (a) action of the driver cell on the first driven cell, (b) action of the driver and of the first driven cell on the second driven cell, (c) action of the driver cell and of the second driven cell on the first driven cell, (d) action of the driven cells on the driver cell. Reprinted with permission from M. Girlanda and M. Macucci, *J. Appl. Phys.* **536**, 92 (2002). Copyright 2002, American Institute of Physics.

The following step consists in solving for the ground state and calculating the electron density of the second driven cell with the inclusion of the effects of the polarization state of both of the first two cells. A second CI calculation is performed considering the electrostatic contributions from the electrons in the two external cells (Fig. 3.15(c)) and then the ground state of the driver cells is computed taking into account the contribution of the electrons of the driven cells as indicated in Fig. 3.15(d).

The steps (b-d) are iterated until the electron densities of two successive interactions converge for all cells. Since the electrostatic contribution between neighboring cells is small, in most cases a single iteration is sufficient to achieve full convergence.

3.9. Results

The intercell distance (distance between cell centers) was set at 442 nm, and polarization of the driver cell was obtained through the application of a a voltage imbalance δV of 10^{-6} V to the gates, as described previously in the text. Under these conditions a full polarization of the driver cell, and consequently of the two driven cells, occurs. In Fig. 3.16(a) the resulting electron densities of the three cells are reported. Since there is no interest in applying a perturbation smaller than 1 μV to the driver cell, a determination of the smallest value of δV that allows full polarization of the cell has not been performed.

The argument that the perturbation needed to polarize the driver cell has an opposite effect on the driven cells and could therefore obstruct the expected alignment of the electrons along the same direction as those in the driver cell, is not completely unfounded, but the net effect is of extremely small magnitude and negligible in comparison with the direct electrostatic action of the electrons in the driver cell.

By including the effect of the discreteness of charge by means a proper quantum mechanical description of the two-electron wave function, we force the number of electrons in a cell to be an integer, as in reality, and therefore we do not introduce artifacts such a the continuous rearrangement of charge that occurs in a simulation with a Thomas-Fermi approximation, a rearrangement that would almost completely screen the the action of the gates.

In other words, due to strong electron localization in the dots, the response of the electron configuration to the applied bias imbalance is extremely sharp, and the electrostatic effect of the electrons localizing along

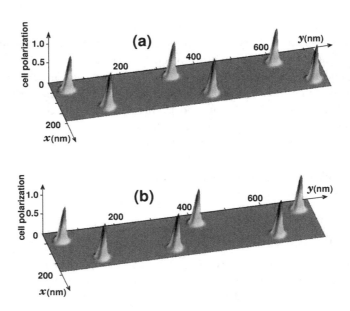

Fig. 3.16.   Electron density in the three active cells for a voltage imbalance $\delta V = 10^{-6}$ V (a) and for a voltage imbalance $\delta V = 10^{-3}$ V (b). Reprinted with permission from M. Girlanda and M. Macucci, *J. Appl. Phys.* **536**, 92 (2002). Copyright 2002, American Institute of Physics.

a diagonal on the polarization of the driver cell is much larger than that arising from the small gate bias imbalance.

It should be noted that the particle-like behavior of the strongly localized electrons is the specific reason that allows charge transfer between dots only in terms of a discrete number of particles: tunneling occurs one "whole" electron at a time. Analogous results can be expected if the cell occupancy is of $4N + 2$ electrons as reported in Ref. 16.

If the gate bias imbalance for the polarization of the driver cell is increased, it may exceed the electrostatic action of the electrons (which remains at a constant value). Thus, for an imbalance $|\delta V|$ larger than a given threshold, the first driven cell will end up in the wrong polarization state, because the action of the driver cell gates will overcome that of the driver cell electrons. Such a situation is sketched in Fig. 3.16(b), which shows the result of an imbalance $\delta V = 1$ mV.

The direct action of the driver cell gates on the driven cells can be easily understood looking at Fig. 3.17. In case (a) an imbalance $|\delta V|$ is applied to the gates and there are no electrons in the driver cell. The response of the electrons of the first driven cell is in this case the result of the asymmetry of the driver cell bias, whose action extends also to the second driven cell, determining for both driven cells a polarization state opposite to that of normal operation. In other words, the more negative gate in the upper right corner of the driver cell pushes away electrons from the upper left dot of the first driven cell. If no electrons are present, this will be the only electrostatic imbalance acting upon the first driven cell, which will polarize in the wrong state, and force into the same state also the second driven cell.

When electrons are placed in the driver cell the normal situation is restored both for the first and the second driven cell (see Fig. 3.17(b)). In

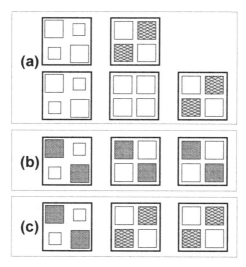

Fig. 3.17. Schematic representation of the opposite action on the driven cells as a result of the gate voltage imbalance (graphically represented with a variation of dot sizes) and of the electrons in the driver cell: (a) if there were no electrons in the driver cell, the electrostatic effect of the voltage imbalance would polarize the first driven cell in the wrong state; (b) if electrons are present in all of the active cells and the bias imbalance is below a threshold, correct polarization is observed in all cells; (c) if electrons are present in all of the active cells and the bias imbalance is too large, correct polarization is observed only in the driver cell, while the other cells are in the opposite polarization state. Reprinted with permission from M. Girlanda and M. Macucci, *J. Appl. Phys.* **536**, 92 (2002). Copyright 2002, American Institute of Physics.

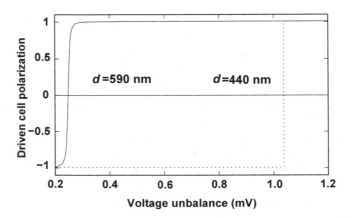

Fig. 3.18.   Polarization of the first driven cell as a function of the gate voltage imbalance
for the driver cell, for two values of the intercell separation: 590 nm (solid curve) and
440 nm (dashed curve). Reprinted with permission from M. Girlanda and M. Macucci,
J. Appl. Phys. **536**, 92 (2002). Copyright 2002, American Institute of Physics.

Fig. 3.17(c) we represent the situation in which too large a value of δV is
applied to the driven cell: this translates into a wrong polarization state in
the first (and consequently in the second) driven cell, for the same reason
as in the case of a driver cell with no electrons.

It is apparent that great care must be taken not to exceed the threshold
value of δV for which an unwanted sign of the polarization of the driven
cell(s) occurs. In Fig. 3.18, the polarization of the first driven cell is reported
as a function of the gate bias imbalance δV, for two different values of the
intercell distance (we refer to the gate layout discussed previously in the
text), and a sign of the applied δV such as to determine a polarization
-1 in the driver cell. In the case of the larger separation (590 nm, solid
line), the applied bias imbalance must be very small to guarantee correct
operation, because a value of just 0.2 mV is already sufficient to perturb
proper operation of the chain. At a smaller intercell separation, instead, δV
needs to reach a value around 1.1 mV before the unwanted polarization,
$+1$, actually occurs.

The dependence of the threshold value of the imbalance δV is shown
in Fig. 3.19. The decay of the critical voltage imbalance with intercell distance is quite rapid, as a result of the screening of the heterostructure surface (Dirichlet boundary conditions have been assumed) on the Coulomb
potential due to the electrons in the driver cell. This because the electro-

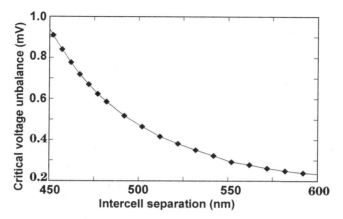

Fig. 3.19. Critical voltage imbalance for failure of the first driven cell as a function of the distance from the driver cell (measured between cell centers).

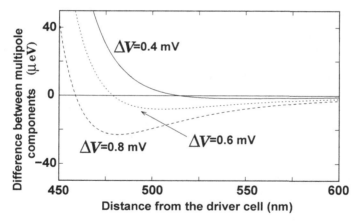

Fig. 3.20. Difference between the multipole component of the potential seen by the electrons in the first driven cell deriving from the electrons in the driver cell and the multipole component deriving from the gate voltage imbalance, as a function of the distance from the driver cell.

static action from the electrons in the driver cell decays faster, as a result of the image charges introduced by the Dirichlet boundary conditions at the surface, than that from the gates, and in particular from the voltage imbalance among the gates of the driver cell. As a consequence, a smaller value of the imbalance δV is sufficient to overcome the effect of the polarization when the distance between the cells is increased. The multipole component due to the gates can be computed separately and compared with the one from the electrons, as a function of the distance from the driver cell. The former can be obtained by subtracting the electrostatic potential due to the gates with $\delta V = 0$ from that with $\delta V \neq 0$, while the latter can be computed by subtracting the electrostatic potential produced by half an electron in each dot from that resulting from two electrons along a diagonal.

The difference between the two components for three different values of the imbalance ($\delta V = 0.4, 0.6, 0.8$ mV) is reported in Fig. 3.20 as a function of the distance from the driver cell. As can be easily concluded, for small values of such a distance the contribution from the electrons is predominant, until the screening caused by the boundary condition at the surface makes it smaller than the contribution from the gates, so that the total multipole component changes sign, determining a change in the polarization state of the driven cell.

References

1. J. H. Davies, *Semicond. Sci. Technol.* **3**, 996 (1993); A. R. Long, J. H. Davies, M. Kinsler, S. Vallis and M. C. Holland, *Semicond. Sci. Technol.* **8**, 1581 (1993); J. H. Davies and I. A. Larkin, *Phys. Rev. B* **49**, 4800 (1994).
2. J. H. Davies, I. A. Larkin, and E. V. Sukhorukov, *J. Appl. Phys.* **77**, 4504 (1995).
3. J. Martorell and D. W. L. Sprung, *Phys. Rev. B* **49**, 13750 (1994).
4. J. Martorell and D. W. L. Sprung, *Phys. Rev. B* **54**, 11386 (1996); D. W. L. Sprung and J. Martorell, *Solid State Comm.* **99**, 701 (1996).
5. M. Governale, M. Macucci, G. Iannaccone, C. Ungarelli, J. Martorell, *J. Appl. Phys.* **85**, 2962 (1999).
6. R. Raichev and P. Debray, *J. Appl. Phys.* **93**, 5422 (2003).
7. D. W. L. Sprung and J. Martorell, *J. Phys. A: Math. Gen.* **30**, 6525 (1997); **31**, 8973 (1998).
8. J. Martorell, Hua Wu and D. W. L. Sprung, *Phys. Rev. B* **50**, 17298 (1994).
9. I. S. Gradshteyn and I. M. Ryzhik, *Table of Integrals, Series and Products*, Academic Press, NY, (1980).
10. L. I. Volkovyskii, G. L. Lunts and I. G. Aramanovitch, *A Collection of Problems on Complex Analysis*, Dover, New York, (1965).
11. J. C. Slater, *Quantum theory of Matter*, McGraw-Hill, New York (1968).
12. R. McWeeny, *Methods of molecular quantum mechanics*, Academic Press, London (1989).
13. S. Gardelis, C. G. Smith, J. Cooper, D. A. Ritchie, E. H. Linfield, and Y. Jin, *Phys. Rev. B* **67**, 033302 (2003).
14. P. Douglas Tougaw and Craig S. Lent, *J. Appl. Phys.* **75**, 1818 (1994).
15. P. Douglas Tougaw, Craig S. Lent and Wolfgang Porod, *J. Appl. Phys.* **74**, 3558 (1993).

16. M. Girlanda, M. Governale, M. Macucci, and G. Iannaccone, *Appl. Phys. Lett.* **75**, 3198 (1999).
17. C. S. Lent, P. D. Tougaw, and W. Porod, *Appl. Phys. Lett.* **62**, 714 (1993).
18. C. S. Lent and P. D. Tougaw, *Proc. IEEE* **85**, 541 (1997).
19. I. Amlani, A. O. Orlov, G. Toth, G. H. Bernstein, C. S. Lent, G. L. Snider, *Science* **284**, 289 (1999).
20. A. O. Orlov, I. Amlani, G. Toth, C. S. Lent, G. H. Bernstein, G. L. Snider, *Appl. Phys. Lett.* **74**, 2875 (1999).
21. C. Single, R. Augke, F. E. Prins, D. A. Wharam, and D. P. Kern, *Semicond. Sci. Technol.* **14**, 1165 (1999).
22. R. P. Coburn, M. E. Welland, *Science* **287**, 1466 (2000).
23. I. I. Yakimenko, I. V. Zozoulenko, K. F. Berggren, *Semicond. Sci. Tech.* **14**, 949 (1999).

CHAPTER 4

Time-Independent Simulation of QCA Circuits

Luca Bonci, Sandro Francaviglia, Mario Gattobigio[a], Carlo Ungarelli[b],
Giuseppe Iannaccone and Massimo Macucci

Dipartimento di Ingegneria dell'Informazione
Università di Pisa
Via Caruso 16, I-56122 Pisa, Italy

4.1. Introduction

The subject of this chapter is the simulation of QCA circuits: we shall discuss several models, along with the main associated results, that we have
developed for treating circuits with different degrees of complexity. As in
the simulation of traditional microelectronic circuits, also in the QCA case
increasing circuit complexity requires the introduction of further degrees
of approximation. While for a single cell or a few coupled cells we have
been able to use rather sophisticated models, such as those based on the
Configuration-Interaction technique, for structures consisting of many cells
the same models become inapplicable, because of the limits imposed by
the maximum computational complexity that we can realistically allow.
We must thus resort to semiclassical approaches, which greatly reduce the
numerical size of the problem, although still yielding results that are consistent with the data from rigorous quantum models (as we have been able
to verify in a few selected cases).

Furthermore, it is important to analyze the behavior of QCA circuits
for nonzero temperatures, in order to determine the maximum operating

[a]Current address: Scuola Normale Superiore, Piazza dei Cavalieri, I-56126 Pisa, Italy
[b]Current address: Physics Department "Enrico Fermi", University of Pisa, Largo Pontecorvo 3, 56127 Pisa, Italy

temperature and the dependence of the error probability on temperature: this cannot be done easily with quantum models, which are best suited to the analysis of the ground state of the system. Classical approaches, instead, can be easily adapted to the investigation of thermal properties, with the application of a few basic concepts from statistical physics.

The issue of thermal behavior is quite important from the point of view of applications, because one of the main limiting factors in the practical usage of most of the proposed nanoelectronic devices is represented by the very low, cryogenic temperatures that are required for their successful operation. Cooling down to liquid helium temperature could only be acceptable for devices able to largely outperform traditional CMOS or to offer qualitatively new opportunities (such as in the case of quantum computing); otherwise the requirement by the industry of room temperature operation is expected to hold firm.

Unfortunately, QCA circuits are very sensitive to thermal fluctuations, even more than single QCA cells, due to the reduced energy separation, as the circuit grows more complex, between the ground state and the first few excited states. As a result, the occupancy of excited states can be nonnegligible even at quite low temperatures. This can result in computational errors, a problem that was first addressed by Lent[4] using entropy considerations.

We want to approach this problem by means of a detailed study of thermal statistics for QCA arrays, with techniques allowing treatment of many cells at the same time. A numerical model will be presented, capable of dealing with relatively short QCA chains in full detail. Moreover, we shall introduce an analytical model, too, which can deal with arbitrarily long chains, if we accept the further approximation of considering only two possible electron configurations in each cell.

Bearing in mind our goal of practical applications, we shall compute both the probability of the system being in the ground state and that of presenting the correct logic output: only one configuration corresponds to the ground state, but the correct logical output is achieved with several other configurations (which will differ for the state of the intermediate cells), thus making the probability of correct logical output larger than that than that of reaching the exact ground state.

Finally, for large and complex circuits we present an approach based on an incomplete, steered exploration of the configuration space, relying on a simulated annealing procedure. Although statistical in nature and therefore sometimes subject to errors, this technique allows the determination of the ground state for relatively big circuits up to about 100 cells.

4.2. Semiclassical Model of QCA Circuits

The basis of our semiclassical approximation consists in assuming the behavior of the electrons as fully classical, except for their ability to tunnel between neighboring cells. This approximation is reasonable if the matrix elements representing the intracell tunneling (i.e. tunneling between the dots in the same cell) are small enough that the electron wave functions can be considered as strongly localized. Such a situation does correspond to the actual conditions in a functioning QCA cells, since high enough barriers are needed to obtain good bistability, as discussed in the previous chapter.

Within each cell we consider two electrons, which can arrange in six possible configurations in the four dots of the cell. The total number of possible configurations is six, because those with two electrons in the same dot are not allowed, due to the excessive electrostatic repulsion. The six possible configurations are shown in Fig. 4.1: for an isolated cell, those with the electrons along a diagonal are degenerate and correspond to the minimum electrostatic energy (since the separation between the electrons is maximum). They are associated with the two logic values 0 and 1. In the presence of neighboring cells, it is possible, particularly in the case of reduced intercell separation, that the other four configurations will be reached, labeled as X and not associated with a logical value, as we shall discuss below.

We start our analysis of QCA circuits with a binary wire, i.e. a linear chain of cells, which propagates the polarization of the first (driver) cell, whose polarization is in turn enforced from the outside, by means of properly biased gates or an otherwise obtained electric field asymmetry. The chain is characterized by just two geometrical parameters: the distance d between the centers of neighboring cells and the distance a between two dots in a cell. In Fig. 4.2, the driver cell is marked with bold lines and only the electron positions are shown. We also consider that a uniformly positive

state 1   state 0 X states

Fig. 4.1.   Possible configurations of a QCA cell with 2 electrons.

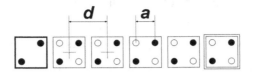

Fig. 4.2.   Geometry of a binary wire.

background charge is distributed among the cells ($e/2$ per dot): this makes
each cell overall neutral, thus preventing anomalous behavior due to the
uncompensated monopole electrostatic field. It is clear that the action of
the uncompensated electrons in a cell would push those of a nearby driven
cell away, thereby preventing the correct propagation of the polarization
state.

We perform calculations for a relative permittivity of 12.9, assuming
that there is no significant permittivity variation between the GaAs and
the AlGaAs layers. In order to keep our treatment as general as possible,
we do not include the effect of the semiconductor-air interface and of the
gate layout defining the dots.[5]

Another possible choice of material system is represented by silicon-
silicon oxide: cells are made up of four silicon dots embedded in silicon
oxide[6]. In silicon cells the electric field lines cross three different materials
with quite different permittivities: silicon, silicon oxide, air. However, most
of them are confined to the silicon oxide region, therefore a good approximation of the behavior of silicon based cells can be obtained from the results
for GaAs by simply replacing the permittivity with that for silicon oxide.
The results for the achievable operating temperature will be conservative,
because part of the field is in the air, which has a unitary permittivity,
therefore in the above mentioned approximation we are underestimating
the actual strength of the electrostatic interaction.

Polarization values are defined,[7] in accordance with the discussion in
the previous chapter, on the basis of the following expression:

$$P = \frac{Q_1 + Q_3 - Q_2 - Q_4}{2},\qquad(1)$$

where Q_i is the charge in the i-th dot, and where we numbered dots starting
from the one at the top right and proceeding counterclockwise. Apart from
configurations with definite -1 or $+1$ polarization, we have those with the
two electrons along one side of the cell, which do not correspond to a logic
state, and which we have previously defined as X states.

In our semiclassical approximation, energy is simply computed as the sum of the electrostatic contributions of a system of point charges, including the term from each electron pair:[8]

$$E = \sum_{i \neq j} \frac{q_i q_j}{4\pi\epsilon_0 \epsilon_r r_{ij}}. \tag{2}$$

In our model, due to the presence of the background charge, we have only two possible values for the total charge in a dot: $+e/2$ or $-e/2$ (since, in order to compensate the negative charge of the two electrons loaded onto each cell, there must be a total positive charge of $2e$, which is uniformly distributed among the four dots). This implies:

$$q_i q_j = \frac{1}{4} e^2 \mathrm{sgn}(q_i q_j). \tag{3}$$

Introducing the ratio $R = d/a$, the total Coulomb energy of a binary wire can be written as

$$E = \frac{e^2}{4a} \frac{1}{4\pi\epsilon_0 \epsilon_r} \sum_{i \neq j} \frac{s_{ij}}{\sqrt{(n_{ij} R + l_{ij})^2 + m_{ij}^2}}, \tag{4}$$

where $n_{ij} \in \{0, \ldots, N_{cell} - 1\}$ is the number of cells between the cell containing dot i and the cell containing dot j, $s_{ij} \in \{-1, 1\}$ indicates the sign of $q_i q_j$, $l_{ij} \in \{-1, 0, 1\}$ and $m_{ij} \in \{0, 1\}$ indicate the position of dots i and j inside the corresponding cells. In particular, l_{ij} is equal to 0 if both dots i and j are on the left side or on the right side of the cell, to -1 if dot i is on the right side and dot j is on the left side and to 1 if dot i is on the left side and dot j is on the right one. Furthermore, m_{ij} is equal to 0 if both dots i and j are at the top or at the bottom of a cell, to 1 if one dot is at the top and the other is at the bottom.

For a general cell layout, with cells located over a 2-D array, the energy E will read

$$E = \frac{e^2}{16\pi a \epsilon_0 \epsilon_r} \sum_{i \neq j} \frac{s_{ij}}{\sqrt{(a_{ij} R + b_{ij})^2 + (c_{ij} R + d_{ij})^2}} \tag{5}$$

where a_{ij} is the horizontal separation, in terms of cells, between dot i and dot j, $b_{i,j} = 0$ if both dots are on the same cell side in the horizontal direction or $b_{i,j} = 1$ if they are on opposite sides. The quantity $c_{i,j}$ is the vertical separation, in terms of cells, between the two dots, and $d_{i,j} = 0$ if both dots are at the top or at the bottom of the corresponding cells.

The energy for a configuration has the following short form

$$E = A g(a) f_c(R), \tag{6}$$

where $A = e^2/(16\pi\epsilon_0\epsilon_r)$ includes the material contribution, $g(a) = 1/a$, i.e. the reciprocal of the interdot distance, and f_c represents the dependence on the geometrical structure of the circuit (the subscript c emphasizes the dependence on the particular configuration).

We have studied a binary wire made up of six cells (five plus the driver cell, whose polarization is fixed). The energy values that correspond to all the 6^5 configurations have been computed, for $a = 40$ nm, and they have been ordered to obtain the energy spectrum, which is shown in Fig. 4.3, expressed in kelvin, with respect to the ground state, (i.e. as the equivalent temperature obtained dividing the energy in joule by the Boltzmann constant), for three different choices of the parameter $R = d/a$. If $R \gg 1$, i.e. $d \gg a$, the interaction between neighboring cells is mainly due to the multipole component, and a discrete spectrum is observed (see Fig. 4.3, case (c))

As R is reduced, the spectrum tends to become continuous (Fig. 4.3, curves (a) and (b)), and the energy of configurations with X states decreases, making them appear also in lower regions of the spectrum.

The first few configurations of some spectra are shown in Fig. 4.4: in the upper plot we have the spectrum for $R = 2.5$, and we observe that the lowest levels include type 1 or 0 states, with a single "kink," i.e. with only one cell flipped with respect to the rest of the chain. As the number of kinks is increased, the increase in energy is roughly proportional. Two energy spectra, for $R = 1.75$ and $R = 2$ are represented in the lower plot:

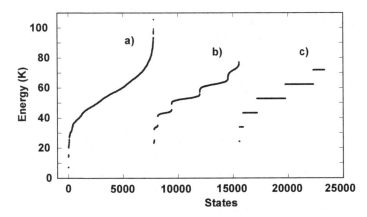

Fig. 4.3.   Energy spectra for different values of R: a) $R = 2$, b) $R = 3.5$, c) $R = 50$.

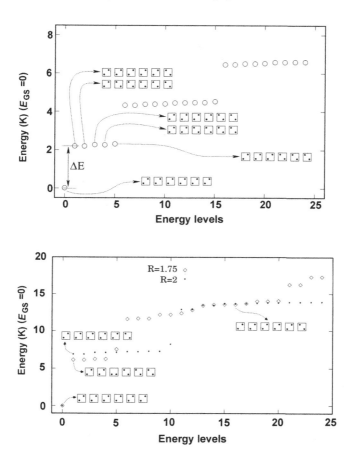

Fig. 4.4. Spectra and low energy configurations for a 6-cell binary wire for $R = 2.5$ (upper plot), and for $R = 2$, $R = 1.75$ (lower plot).

in the case of $R = 1.75$ the first excited level includes an X state and the energy split is lower than in the case with $R = 2$. For smaller values of R, a continuous spectrum is approached, with a progressive merging of the plateaus: in particular, if we decrease R while keeping a constant, thereby reducing the separation between neighboring cells, the difference between the energy of the ground state and that of the first excited state are expected to increase as $1/R$, due to the increased electrostatic interaction. This is however true only down to a threshold value of R, below which the splitting between the first excited state and the ground state starts decreasing, as illustrated in Fig. 4.5, where the energy split is plotted as a function of R

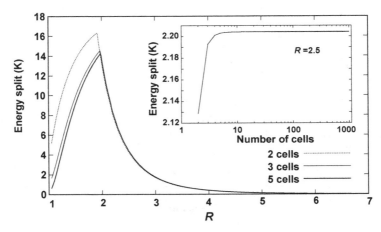

Fig. 4.5. Energy splitting as a function of R between the first excited state and the ground state for a binary wire with 2 (dotted line), 3 (dashed line), and 5 (solid line) driven cells; the inset contains a plot of the energy splitting for $R = 2.5$ as a function of the number of cells. Adapted with permission from C. Ungarelli, S. Francaviglia, M. Macucci, G. Iannaccone, *J. Appl. Phys.* **87**, 7320 (2000). Copyright 2000 American Institute of Physics.

for a cell size a of 40 nm, for a wire with 2 (dotted line), 3 (dashed line), and 5 (solid line) cells. This change of behavior can be explained on the basis of the results discussed previously: below the threshold value for R, the configuration for the first excited state includes a cell in the X state, thus altering the logic operation of the wire and reducing the energy difference between the first excited state and the ground state. The splitting between the first excited state and the ground state is reported as a function of the number of cells in the inset of Fig. 4.5. These results have been obtained for $R = 2.5$, a value of R for which no X state appears. If the number of cells increases, the splitting reaches a constant saturation value. This can be explained by considering that the strength of the electrostatic interaction drops sharply along the chain, and thus the energy of the first excited state (the one with a cell at the end of the wire polarized in an opposite way compared to the others), does not significantly change when cells are added at the other end of the chain. The energy splitting is inversely proportional to a (due to the inverse dependence on distance of the Coulomb interaction) and thus can be increased by scaling down cell dimensions.

A calculation of the energy splitting can be performed for a generic QCA gate, in particular we show results for a majority voting gate (represented

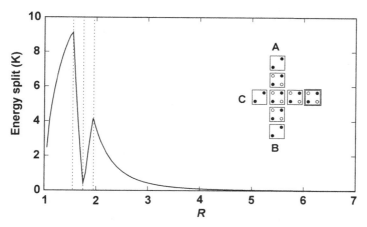

Fig. 4.6. Energy splitting as a function of R between the first excited state and the ground state for a majority voting gate (whose layout is represented in the inset).

in the inset of Fig. 4.6) with the choice of inputs $A = 1$, $B = 1$, $C = 1$. In this case two local maxima and a local minimum occur before $\Delta E(R)$ becomes proportional to $1/R$, as marked with dotted lines in Fig. 4.6.

4.3. Thermal Behavior

Let us introduce the partition function for a binary wire:

$$Z = \sum_i e^{-\beta E_i}. \tag{7}$$

The variable E_i represents the energy associated with the i-th configuration, $\beta = kT$, k is the Boltzmann constant and T the absolute temperature. The summation is performed over all possible configurations with the first cell kept in a constant logic state. The probability P_{gs} that the wire is in the ground state is computed as the ratio of the Boltzmann factor for the ground state to the partition function:

$$P_{gs} = \frac{e^{-\beta E_{gs}}}{Z} = \frac{1}{1 + \sum_{i \neq gs} e^{-\beta \Delta E_i}}, \tag{8}$$

where $\Delta E_i = E_i - E_{gs}$, and the summation runs over all excited states.

The probability P_{gs} is not the only quantity of interest: we are also interested in the probability P_{clo} of achieving the correct logic output (consistent with the applied logic input), which has a value larger than P_{gs}, due to the fact that not only the ground state, but several different configurations do include the expected polarization of the output cell. The probability P_{clo} can be evaluated summing the probabilities relative to all the configurations sharing the correct output value, labeled with the subscript j:

$$P_{clo} = \frac{\sum_j e^{-\beta E_j}}{Z} . \tag{9}$$

For a six-cell chain we have computed P_{gs} and P_{clo} as a function of the ratio of ΔE (energy splitting between the ground and the first excited state) to kT and the results are shown in Fig. 4.7: for temperatures that are large compared to the energy splitting, i.e. $\Delta E/(kT) \ll 1$, all the configurations (a total of 6^5) tend to be equally probable, thus the probability P_{gs} attains its minimum value $1/6^{N-1}$. The probability of correct logic output has instead a minimum value of $1/6$, the obvious consequence of the six possible states of the output cell being present with the same probability.

Let us point out that even if an error probability of a few percent may not seem acceptable for practical applications, the actual situation is different: data readout must be performed with some type of detector,[5] whose time constant is unavoidably longer (often much longer) than the typical time a QCA circuit takes to reach the ground state. Each reading will thus

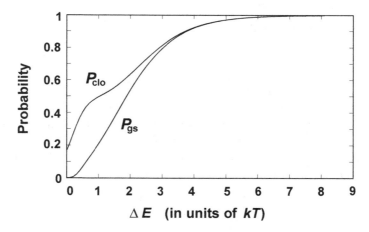

Fig. 4.7. Probability of being in the ground state P_{gs} and of achieving the correct ground state P_{clo} for a binary wire, as a function of the energy splitting ΔE between the first excited state and the ground state. Adapted with permission from C. Ungarelli, S. Francaviglia, M. Macucci, G. Iannaccone, *J. Appl. Phys.* **87**, 7320 (2000). Copyright 2000 American Institute of Physics.

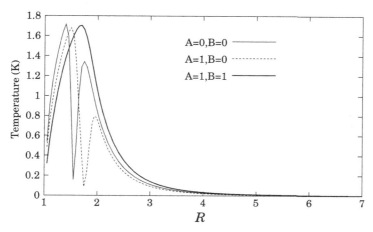

Fig. 4.8. Maximum temperature at which $P_{gs} = 0.9$ for a majority voting gate, as a function of the ratio R.

be represented by an average value, to be compared with a decision threshold: in most cases, this will result in an error probability, at the output of the readout system, much smaller than that associated with an instantaneous measurement.

We can compute the maximum operating temperature for a given circuit, i.e. the temperature at which the probability of correct logical output reaches a threshold value. Let us, for example, compute the temperature at which P_{gs} or P_{clo} equal 0.9 as a function of the ratio R for a majority voting gate: results are shown in Fig. 4.8 (for P_{gs}) and in Fig. 4.9 (for P_{clo}), for different choices of the input variables. The thin solid line corresponds to the maximum operating temperature for a majority voting gate with $A = 0$, $B = 0$, the dashed line for $A = 1$, $B = 0$, and the thick solid line for $A = 1$, $B = 1$. For the input condition $A = 0$, $B = 0$ the correct logical output would be 0, but if R is less than a threshold, the ground state and the first excited state do not have the output cell in the correct configuration, therefore it is not possible to get a nonzero P_{clo} at any temperature. Only for a large enough R, the ground state changes and starts corresponding to the correct logical output, so that we can have P_{clo} over 90% at a finite temperature.

We can also study the maximum operating temperature (again for $P_{clo} = 0.9$) as a function of the interdot distance a for a constant value of $R = 2.5$. In Fig. 4.10 we report such a result not only for the Gallium

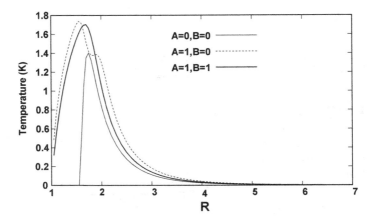

Fig. 4.9. Maximum temperature at which $P_{clo} = 0.9$ for a majority voting gate, as a function of the ratio R.

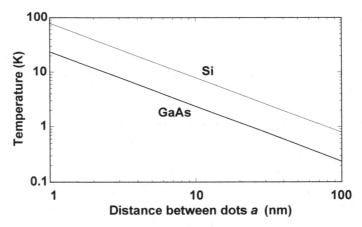

Fig. 4.10. Maximum operating temperature as a function of distance a for the Si and GaAs material systems.

Arsenide material system (thick line) that we have considered so far in this chapter, but also for the Si-SiO$_2$ material system (thin line), assuming, in this latter case, the permittivity equal to that of silicon oxide ($\varepsilon_r = 3.9$), since most of the electric field lines are located in silicon oxide, as already discussed. Due to the smaller permittivity for silicon, the electrostatic interaction is increased, and this allows proper operation at somewhat higher temperature values with respect to the GaAs implementation.

From the computational point of view, we need to point out that, since the number of possible configurations for a circuit with N cells (one of which is fixed, being the driver cell) is 6^{N-1}, the CPU time required for the computation of the energy of all configurations grows exponentially with N. This limits the length of the binary wires that can be treated with this approach to about ten cells, if we want to obtain results in a reasonable time. A somewhat larger number of cells can be treated considering only two possible configurations per cell, the ones corresponding to just the two logical values 0 and 1 and therefore neglecting all the X states. This can be done if the ratio R of the intercell separation to the interdot spacing is large enough: as we have already discussed, X states start appearing as a result of cells too close to each other, while otherwise they are associated with too large energies and are thus usually unoccupied. With this further approximation, circuits up to about 20 cells can be treated with success, as long as the temperatures considered are low enough that X states will not be excited.

4.4. Analytical Model

We can extend our investigation of the thermal behavior of binary wires beyond the 20-30 cell limit associated with numerical calculations by developing an analytical treatment,[9] which, for the particular case of the linear chain, is relatively straightforward, if only the configurations corresponding to the logic states 0 and 1 are included. Let us start by considering a generic 1-dimensional chain made up of N cells and introduce the following Ising Hamiltonian

$$\mathcal{H} = -J \sum_{i=1}^{N-1} \sigma_i \sigma_{i+1} \,, \tag{10}$$

where i is the cell index and σ_i represents the polarization (+1 or −1) of the i-th cell, J is one half of the splitting ΔE between the ground state and the first excited state.

In the following, we shall assume that the polarization of the driver cell is 1, so that the ground state is the one with $\{\sigma_i = 1, \forall i\}$. The partition function of the N-cell system, based on the Hamiltonian (10), will thus read

$$Z = \sum_{\{\sigma\}} e^{-\beta \mathcal{H}} \,, \tag{11}$$

where $\{\sigma\}$ indicates summation over all possible states, $\{\sigma_1 = 1, \sigma_i =$

$\pm 1 , i = 2 \ldots N\}$. Equation (11) can be rewritten as

$$Z = \sum_{\sigma_2 \ldots \sigma_N} V(1, \sigma_2) \, V(\sigma_2, \sigma_3) \ldots V(\sigma_{N-1}, \sigma_N) , \qquad (12)$$

with $V(\sigma, \sigma') = e^{\beta \sigma \sigma'}$. The evaluation of the r.h.s. of Eq. (12) can be performed introducing the transfer matrix [10]

$$\mathcal{V} = \begin{pmatrix} e^{\beta J} & e^{-\beta J} \\ e^{-\beta J} & e^{\beta J} \end{pmatrix} , \qquad (13)$$

with eigenvalues

$$\lambda_+ = e^{\beta J} + e^{-\beta J} , \qquad \lambda_- = e^{\beta J} - e^{-\beta J} . \qquad (14)$$

Thus we have

$$\mathcal{V}^{N-1} = \begin{pmatrix} \frac{\lambda_+^{N-1} + \lambda_-^{N-1}}{2} & \frac{\lambda_+^{N-1} - \lambda_-^{N-1}}{2} \\ \frac{\lambda_+^{N-1} - \lambda_-^{N-1}}{2} & \frac{\lambda_+^{N-1} + \lambda_-^{N-1}}{2} \end{pmatrix} . \qquad (15)$$

The partition function (12) therefore reads

$$Z = [\mathcal{V}^{N-1}]_{11} + [\mathcal{V}^{N-1}]_{12} = (e^{\beta J} + e^{-\beta J})^{N-1} , \qquad (16)$$

where the subscripts indicate elements of the $\mathcal{V}^{N-1}$ matrix. Via this expression, it is possible to derive an analytical relationship for the probability of the system being in its ground state, as a function of temperature and of the energy splitting between the first excited state and the ground state. The ground state for the Hamiltonian in Eq. (10) is $E_{gs} = -J(N-1)$, thus we have

$$P_{gs} = \frac{e^{-\beta E_{gs}}}{Z} = \frac{e^{\beta J(N-1)}}{Z} = \frac{1}{(1 + e^{-\beta \Delta E})^{N-1}} . \qquad (17)$$

We can now obtain an expression also for the probability of correct logical output. To this purpose we need to compute the occupation probability of all states with $\sigma_1 = \sigma_N = 1$. This can be achieved by defining a "reduced" partition function:

$$Z_R = \sum_{\sigma_2 \ldots \sigma_{N-1}} V(1, \sigma_2) \, V(\sigma_2, \sigma_1) \ldots V(\sigma_{N-1}, 1) , \qquad (18)$$

where again $V(\sigma, \sigma') = e^{\beta \sigma \sigma'}$. Using the transfer matrix (13), it follows that $Z_R = [\mathcal{V}^{N-1}]_{11}$, and hence

$$P_{clo} = \frac{Z_R}{Z} = \frac{[\mathcal{V}^{N-1}]_{11}}{[\mathcal{V}^{N-1}]_{11} + [\mathcal{V}^{N-1}]_{12}} = \frac{1}{2} \left[1 + (\tanh(\beta \Delta E/2))^{N-1} \right] . \qquad (19)$$

By means of these analytical expressions P_{gs} and P_{clo} have been computed as a function of temperature for a 6 cell chain, with cell size $a = 40$ nm and cell separation $d = 100$ nm. In Fig. 4.11 we compare the analytical results (dashed lines) with the numerical ones (solid lines). The agreement with the more detailed numerical results (obtained including all six possible configurations) is almost perfect for low temperature values (below about 2 K, corresponding to low error probability). At higher temperatures, due to the relevance of X states, which have not been included in the analytical approach, some differences start appearing. In particular, it is clear that for large temperature values the analytical approach cannot reproduce the $1/6$ value for P_{clo}, due to the inclusion of only two possible states, and, indeed, the analytical P_{clo} approaches $1/2$. A similar argument can be made for P_{gs}, which goes down only to $1/2^5$ in the analytical case, while it reaches an extremely small value $(1/6^5)$ in the numerical calculations with all six states included.

The probabilities P_{clo} and P_{gs} as a function of the ratio $\Delta E/(kT)$ are reported in a semilogarithmic scale in Fig. 4.12. Analytical model results are reported with a thick solid line, while numerical model results are shown with a thin solid line $(R = 2)$, a dashed line $(R = 2.5)$, and a dotted line $(R = 4)$. The agreement gets better as R increases, confirming the

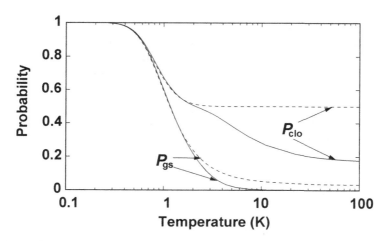

Fig. 4.11. Comparison between the analytical (solid line) and numerical (dashed line) results for P_{gs} and P_{clo}, for a six cell chain, as a function of temperature. Adapted with permission from C. Ungarelli, S. Francaviglia, M. Macucci, G. Iannaccone, *J. Appl. Phys.* **87**, 7320 (2000). Copyright 2000 American Institute of Physics.

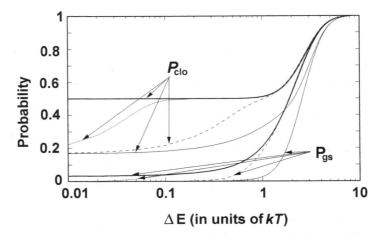

Fig. 4.12.   Probabilities P_{clo} and P_{gs} as a function of the energy splitting between the first excited state and the ground state, for $R = 2$ (thin solid lines), $R = 2.5$ (dashed lines), $R = 4$ (dotted lines). Adapted with permission from C. Ungarelli, S. Francaviglia, M. Macucci, G. Iannaccone, *J. Appl. Phys.* **87**, 7320 (2000). Copyright 2000 American Institute of Physics.

decreasing relevance of the X states. For $\Delta E/(kT)$ significantly greater than 1, the error probability is quite small, and the analytical approach is very reliable.

The analytical expression is particularly useful to provide estimates of the maximum operating temperature for a QCA chain as a function of the number of cells. The results for the maximum operating temperature allowing a given correct logic output probability, are presented in Fig. 4.13 versus the number of cells. The various lines correspond to different values of the probability of correct logical output: $P_{clo} = 0.6$ (solid line), 0.9 (dashed line), 0.99 (dotted line), for a cell size $a = 40$ nm and an intercell separation $d = 100$ nm. We have found a logarithmic decrease of the maximum operating temperature with cell number if such a number is above a few tens, confirming the results of Ref. [4]. This appears as a straight section in the logarithmic representation of Fig. 4.13.

4.4.1. *Numerical simulation of more complex circuits*

Circuits that are more complex than a binary wire can be treated with our numerical approach: an example is shown in Fig. 4.14(a): this layout implements the logic function $AB + CD$ with a total of 18 cells, 7 of which

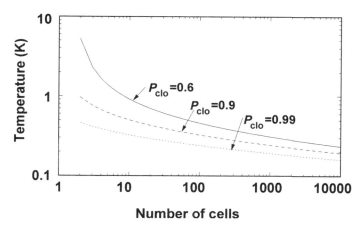

Fig. 4.13. Maximum operating temperature of a QCA binary wire as a function of the number of cells in the wire, for different values of P_{clo}. Adapted with permission from C. Ungarelli, S. Francaviglia, M. Macucci, G. Iannaccone, *J. Appl. Phys.* **87**, 7320 (2000). Copyright 2000 American Institute of Physics.

are driver cells (i.e. with logic levels enforced from the outside). Therefore only 11 cells can span the configuration space, which will consist of 6^{11} configurations. In Fig. 4.14(b) we report results for the ground state probability P_{gs} and the correct logical output probability P_{clo} for the input state $A = B = C = D = 1$, while in Fig. 4.14(c) the same quantities are shown for the input state $A = 0, B = C = D = 1$. The difference between the results for the two choices of input vectors is due to the variation of the energy splitting between the first excited state and the ground state: for $A = B = C = D = 1$ such a splitting is 0.159 meV, while for $A = 0, B = C = D = 1$ it decreases down to 0.086 meV.

The main limitation to treatment of large circuits results from the exponentially growing dimension of the configuration space: even considering only two states per cell, circuits with more than 20-25 cells cannot be conveniently treated. For larger structures, it is however possible to resort to a different approach, which does not require a complete exploration of the configuration space and relies instead on a simulated annealing procedure to reach the ground state configuration.

We start the calculation with the system in a randomly selected configuration of energy E_0 at an initial temperature T_0. If the configuration energy at the current iteration is indicated with E_a, at the beginning $E_a = E_0$. A new configuration can be chosen at random by simply moving one electron

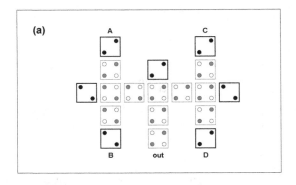

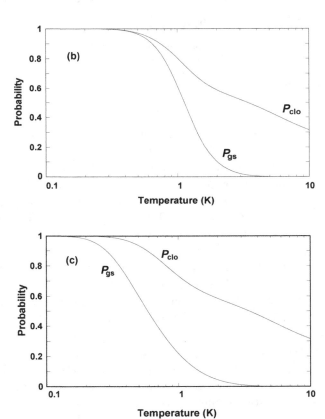

Fig. 4.14. Layout of a four-input QCA circuit (a), and ground state and correct logical output probabilities as a function of temperature for $A = B = C = D = 1$ (b) and for $A = 0$, $B = C = D = 1$.

between two nearby dots within the same cell. The new configuration is accepted with a probability p_{new} depending on its energy E_{new} and on the energy E_a of the current configuration:

$$p_{new} = \begin{cases} 1 & \text{if } E_{new} \leq E_a \\ \exp[-(E_{new} - E_a)/kT] & \text{if } E_{new} > E_a \end{cases} . \qquad (20)$$

These two conditions warrant that the system evolution proceeds along trajectories of descending energy (first expression), and that the system does not get stuck in metastable states, that is local energy minima (the nonzero probability of moving to a higher energy state allows to leave such minima). The procedure is iterated many times, while progressively reducing the temperature, until a stable configuration is obtained, which should correspond to the ground state.

In Fig. 4.15 an evolution to the ground state in shown: for a binary wire of six cells, the instantaneous configuration energy and the temperature are plotted as a function of the iteration number.

In terms of the required computational resources, simulated annealing is increasingly preferable to the complete exploration of the configuration space as the system size grows. The possibility that the simulated annealing procedure may get stuck in a configuration corresponding to a local energy minimum cannot however be neglected. An example in the case of the two-to-one multiplexer is shown in the inset of Fig. 4.16, the probability that

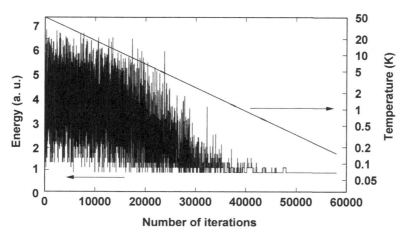

Fig. 4.15.   Evolution of a QCA circuit toward the ground state in a simulated annealing calculation.

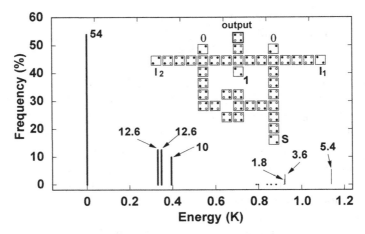

Fig. 4.16.   Relative frequency of the outcomes of simulated annealing calculations for the circuit shown in the inset.

the simulated annealing converges to the ground state is only 0.54, if we start form a random configuration. Other low-lying excited states, their energy, and the corresponding probability of being obtained as a result of the annealing procedure are shown in Fig. 4.16. A safer approach consists in performing a sort of "thermal cycling," that consists in several cycles in which the system is allowed to reach a stable state, then the temperature is brought back to a higher value, forcing the system back into an excited state, and then slowly reduced again to relax the system into another stable configuration. It is possible to show that among all of the stable states obtained, the probability to reach the ground state is $1 - (1 - P_1)^m$, where m is the number of thermal cycles and P_1 is the probability of obtaining the ground state without thermal cycling. Thus, for the particular example of the multiplexer, in which $P_1 = 0.54$, the probability of finding the ground state with eight thermal cycles would be 0.998.

References

1. C. S. Lent, P. D. Tougaw, and W. Porod, *Appl. Phys. Lett.* **62**, 714 (1993).
2. M. Governale, M. Macucci, G. Iannaccone, C. Ungarelli, J. Martorell, *J. Appl. Phys.*, **85**, 2962 (1999).
3. P. D. Tougaw and C. S. Lent, *J. Appl. Phys.* **75**, 1818 (1994).
4. C. S. Lent, P. D. Tougaw, and W. Porod, in the *Proceedings of the Workshop on Physics and Computing*, Nov. 17-20, 1994, Dallas, TX, p. 1.

5. G. Iannaccone, C. Ungarelli, M. Macucci, E. Amirante, M. Governale, *Thin Solid Films*, **336**, 145 (1998).

6. F. E. Prins, C. Single, F. Zhou, H. Heidemeyer, D. P. Kern, E. Plies, *Nanotechnology*, **10**, 132 (1999).

7. M. Girlanda, M. Governale, M. Macucci, G. Iannaccone, *Appl. Phys. Lett.*, **75**, 3198 (1999).

8. J. D. Jackson, *Classical Electrodynamics* Wiley, New York (1962), p. 20.

9. C. Ungarelli, S. Francaviglia, M. Macucci, G. Iannaccone, *J. Appl. Phys.* **87**, 7320 (2000).

10. See e.g. J. R. Baxter, *Exactly solved models in statistical mechanics*, Academic Press, London (1982).

CHAPTER 5

Simulation of the Time-Dependent Behavior of QCA Circuits with the Occupation-Number Hamiltonian

Irina Yakimenko and Karl-Fredrik Berggren

Department of Physics and Measurement Technology (IFM)
Linkoeping University
SE-58183 Linkoeping, Sweden

5.1. Introduction

In this chapter we will describe the results of the investigation of the dynamical response in model quantum cellular automata, based on the Hubbard Hamiltonian approach, with the subsequent numerical simulation of the time evolution of polarization for a chain of QCA cells without and with dissipation.[1] We will also discuss the influence of imperfections or stray charges on the dynamic evolution and switching properties of QCA. A numerical analysis has been performed to study in detail the dependence of the switching properties of the model QCA on irregularities in the spatial distribution of the cells, on the local variations of the interdot barriers, and on the presence of stray charges. The results of this analysis demonstrate that the dynamic switching of the long chains of quantum cells is very sensitive to the presence of perturbations.[2]

5.2. Modeling of Chains of Quantum Cells

Let us consider a chain of QCA cells consisting of a driver cell on the left end and of a number of cells along the chain, interacting electrostatically. Each cell is formed by four dots in the corners of a square with a separation a. The cells are occupied by two electrons which may tunnel between neighboring dots. The electronic configuration in each dot depends on the particular cellular quantum states. Intercell tunneling is assumed to be totally suppressed, which means that no electron transfer between cells occurs (in practice the distance between neighboring cells should exceed at least twice the distance between dots in a cell). Due to the absence of charge flow, low power consumption can be expected.

Suppose that each cell is described by the Hamiltonian of the extended Hubbard-type,[3]

$$H = \sum_{m,\sigma}(E_{0,m} + V_m)\hat{n}_{m,\sigma} - L\sum_{m,\sigma}(\hat{a}^{\dagger}_{m+1,\sigma}\hat{a}_{m,\sigma} + \hat{a}^{\dagger}_{m,\sigma}\hat{a}_{m+1,\sigma})$$

$$+ \sum_m E_Q\hat{n}_{m,\uparrow}\hat{n}_{m,\downarrow} + \sum_{m>j,\sigma,\sigma'} V_Q\frac{\hat{n}_{m,\sigma}\hat{n}_{j,\sigma'}}{|R_m - R_j|}. \tag{1}$$

Here we use the standard second-quantized notation, where the operator $\hat{a}_{m,\sigma}(\hat{a}^{\dagger}_{m,\sigma})$ annihilates (creates) an electron in dot m with spin σ and $\hat{n}_{m,\sigma}$ is the number operator for the mth dot. The first term in Eq. (1) is an on-site energy, where V_m represents the potential energy of an electron in dot m due to charges in other cells,[3] and $E_{0,m}$ represents the on-site energy at dot m defined by the other sources.

We will consider the case in which $E_{0,m}$ is equal for all dots except for dots in the driver cell. Then $E_{0,m} = E_0$, and we let $E_0 = 0$ be the reference energy. The second term in Eq. (1) represents electron tunneling between nearest-neighbor dots, described by the tunnel coupling coefficients L. The third term is the Hubbard energy, E_Q, i.e. the energy required to put two electrons with opposite spins in the same dot. The last term corresponds to the Coulomb interaction between electrons in two different dots ($|R_m - R_j|$ is the distance between dots m and j in units of a).

In order to estimate the parameters L, E_Q, and V_Q, we have used a simplified model for the quantum dots which are fabricated from heterostructures of $GaAs/Al_xGa_{1-x}As$. In the limit of a strong confinement in the direction perpendicular to the interface, a quantum dot may be regarded as a two-dimensional system. Hence the potential associated with the bare electrostatic confinement may be assumed of the form

$$V(x,y) = \frac{1}{2}m^*\omega_x^2x^2 + \frac{1}{2}m^*\omega_y^2y^2 \tag{2}$$

where $\hbar\omega_x$ and $\hbar\omega_y$ define the steepness of the confinement in the lateral x- and y-directions. The wave function for the first energy level of a dot is

$$\psi(x,y) = \left(\frac{\alpha_1\alpha_2}{\pi}\right)^{1/2} \exp\left[-\frac{1}{2}(\alpha_1^2x^2 + \alpha_2^2y^2)\right]. \tag{3}$$

To get a first rough estimate of the bare confining potential for the four dots in a cell, we divide the cell into four equal quadrants. In each of these quadrants the potential is given by $V(x-X, y-Y)$ with (X, Y) denoting the center of the parabola. The potential is shown schematically in Fig. 5.1(a). Using perturbation theory we may now calculate the parameters of the cell. When evaluating the integrals defining L, E_Q, and V_Q we have used numerical integration.

Figure 5.1(b) displays the Coulomb repulsion energy and Fig. 5.1(c) the coupling coefficient between nearest-neighboring dots within one cell as a function of the distance, for the confinement potential $\hbar\omega_x = \hbar\omega_y = 18$ meV. Strong confinement would be needed for operation at room temperature. For the present type of system, the Coulomb energy is of the order of 1 meV. Figure 5.1 shows that the ratio V_Q/L is in the range from 48 to 1200 for an inter-dot distance from 32 nm to 50 nm. For an inter-dot distance smaller than 32 nm the electrons do not localize on the individual dots. In our calculations the inter-dot distance has been chosen as 36 nm for the chains with two and three cells, and as 34 nm for the chain with six cells. Furthermore, we have used the value 13.1 for the dielectric constant and $0.067\, m_e$ for the effective mass. The corresponding parameters L, V_Q and E_Q are as follows: $L \simeq 0.038$ meV, $V_Q \simeq 1.7$ meV, and $E_Q \simeq 8.53$ meV.

5.3. Time Evolution of Polarization for a Chain of QCA Cells without Dissipation

For each cell, we have to solve the time-dependent Schrödinger equation,

$$i\hbar \frac{\partial |\Psi(t)\rangle}{\partial t} = H|\Psi(t)\rangle. \tag{4}$$

For this purpose the wave function $|\Psi(t)\rangle$ may be expanded in a complete set of two-electron determinants $\{|k, l\rangle\}$. If the electrons are assumed to have opposite spins, then $k, l = 1, 2, 3, 4$ and we write

$$|\Psi(t)\rangle = \sum_{k,l}^{4} c_{k,l}(t)|k, l\rangle. \tag{5}$$

Here $c_{k,l}(t)$ are expansion coefficients and the site representation $|k, l\rangle$ refers to the situation that if one electron occupies dot k, the other one resides in dot l with opposite spin. When we substitute Eq. (5) into Eq. (4) with the Hamiltonian given by Eq. (1), we obtain the time-dependent equation for $c_{k,l}$,

$$\hbar \frac{\partial c_{k,l}}{\partial t} = iL(c_{k,l-1} + c_{k,l+1} + c_{k-1,l} + c_{k+1,l})$$
$$- i\delta_{k,l}E_Q c_{k,l} - i(V_p(1 - \delta_{k,l}) + V_k + V_l)c_{k,l}. \tag{6}$$

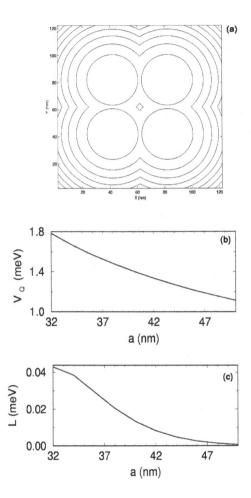

Fig. 5.1.   (a) Schematic diagram of the bare confining potential in a cell. The strength
of the confinement in the x and y directions for each dot is chosen as $\hbar\omega_x = \hbar\omega_y =$
18 meV. The distance between nearest-neighboring dots is $a = 36$ nm; (b) the Coulomb
energy between two electrons residing in nearest-neighboring dots within one cell and
(c) the coupling coefficient between nearest-neighboring dots as a function of the inter-
dots distance for $\hbar\omega_x = \hbar\omega_y = 18$ meV. Reprinted with permission from C.-K. Wang,
I. I. Yakimenko, I. V. Zozoulenko, and K.-F. Berggren, *J. Appl. Phys.* **84**, 2684 (1998).
Copyright 1998, American Institute of Physics.

Here $V_p = V_Q$ for the Coulomb coupling between nearest-neighbor dots and $V_p = V_Q/\sqrt{2}$ for antipodal dots within one cell; $c_{k,0} = c_{k,4}, c_{k,5} = c_{k,1}$ and $c_{0,k} = c_{4,k}, c_{5,k} = c_{1,k}$. Solving this set of temporal equations with given initial conditions, we can study the evolution of the polarization in the different cells. The polarization of the cell is defined by

$$P(t) = \frac{\rho_1(t) + \rho_3(t) - \rho_2(t) - \rho_4(t)}{\rho_1(t) + \rho_3(t) + \rho_2(t) + \rho_4(t)}, \tag{7}$$

where $\rho_m(t)$ is the expectation value of the number operator for dot m,

$$\rho_m(t) = \langle \Psi(t)|\hat{n}_m|\Psi(t)\rangle. \tag{8}$$

First, we consider the case when relaxation processes are not included, that is, the case that has been already studied by Tougaw and Lent.[4] In addition, we will investigate how the evolution depends on the choice of the time τ over which switching of the driver cell is completed. In the way we solve the dynamics, we allow the system to remain instantaneously in its ground state when τ is chosen sufficiently long; in rough terms, one should expect $\tau \gg \hbar/\Delta E = \tau_{ex}$ where ΔE is the level spacing.

For a chain of two cells, Fig. 5.2 shows the time evolution of the polarization of the second cell for some different switch times. As an example, we choose the initial condition for the driver cell as $P_1 = 1$. The enforced polarization change is chosen to be linear in time. As shown by Fig. 5.2, the polarization of the second cell will follow that of the driver more closely as the switch time is increased. With longer switch times, the amplitude of the quantum oscillations decreases and the driven cell effectively switches to a new state of polarization with $P_2 \simeq -1$. For short switch times, however, the second cell will never reach the desired polarization, but will continue to oscillate. For this abrupt change, the second cell obviously does not follow the driver cell. The second cell is excited because the switch time is shorter than the time τ_{ex} associated with the transition to an excited state. For the present case this occurs for $\tau_{ex} \sim 30$ ps. On the other hand, for a very long switch time the second cell follows the first cell closely, which indicates that it remains in its ground state at all instants. In this case the switching is thus adiabatic.

In addition to the coarse arguments above, one may also estimate the time τ_{adiab} required for the adiabatic switching of the four-dot cell array by employing time-dependent perturbation theory. A quantum system subject to the external time-dependent potential $V(t)$ will remain in its ground

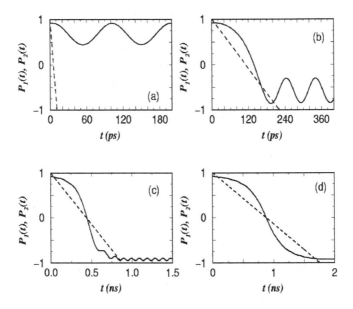

Fig. 5.2. Time evolution of the polarization for a chain consisting of two cells. The polarization $P_1(t)$ of the driver cell is switched linearly (dashed line) and the polarization $P_2(t)$ of the second responding cell is shown by the solid line. The different cases correspond to different switch times of the driver cell: (a) $\tau = 10$ ps; (b) $\tau = 220$ ps; (c) $\tau = 0.90$ ns, (d) $\tau = 1.75$ ns. Reprinted with permission from C.-K. Wang, I. I. Yakimenko, I. V. Zozoulenko, and K.-F. Berggren, *J. Appl. Phys.* **84**, 2684 (1998). Copyright 1998, American Institute of Physics.

state if the potential changes slowly enough,

$$|\omega_{nm}^{-1}\frac{d}{dt}\langle n|V(t)|m\rangle| \ll |E_n - E_m|; \ \omega_{nm} = \frac{E_n - E_m}{\hbar}. \qquad (9)$$

In the system of four dots the energy levels E_n change as the potential of the driver cell, $V(t)$, varies. The minimum energy difference between the ground and the first excited state, $\Delta^{\min} \approx 0.01$ meV, corresponds to zero polarization of the driver cell, $P_1 = 0$. Using this in Eq. (9) one gets an estimate of the maximum time needed for adiabatic switching, $\tau_{\text{adiab}}^{\max} \approx 210$ ps. When the polarization of the driver cell $P_1 = 1$, similar arguments give $\Delta^{\max} \approx 0.04$ meV and $\tau_{\text{adiab}}^{\min} \approx 12$ ps. The calculations show that the energy separation between E_1 and E_2 changes linearly as the polarization of the driver cell is varied from zero to one. Thus, one roughly gets $\tau_{\text{adiab}} \approx$

$(\tau_{\text{adiab}}^{\max} + \tau_{\text{adiab}}^{\min})/2 = 110$ ps. This value of the time required for the adiabatic switching of the cell is in good agreement with the results of the numerical simulations, see Fig. 5.2.

Next, we turn to the case of a linear chain with three cells. The results for different switch times of the driver cell are shown in Fig. 5.3, which clearly demonstrates that the polarizations of the driven cells do not match that of the driver cell for reasonable choices of τ. However, the amplitude of the oscillations for the polarization of the second cell is weaker than that for the third cell. In addition, the second cell follows the driver cell more closely than the third one does. For QCA implementation, the response of the third cell is unsatisfactory due to the nonadiabatic evolution of the second cell. Hence the system stays in a mixed state. This can be clearly seen in the energy representation, which shows that the occupation probabilities of higher energy states are non-zero. The results for three cells thus demonstrate the fact that QCA can be hardly operable in the adiabatic mode discussed here. Some other operational modes of a QCA have been proposed,[5,6] requiring that the polarization of the driver cell and the inter-dot barriers in the responding cells have to be changed simultaneously.

5.4. Time Evolution of Polarization for a Chain of QCA Cells with Dissipation

To include inelastic processes into the system dynamics we will use a phenomenological approach, and for this purpose it is convenient to introduce the energy representation. For each cell, we expand the wave function $|\Psi(t)\rangle$ into a complete set of eigenstates $|\alpha\rangle_t$ for the two electrons with opposite spins occupying the four dots in a cell. Now suppose that this cell is subject to an external field that varies with time in accordance with a prescribed change of the polarization of the driver cell. In the energy representation, we may then solve the eigenvalue equation

$$H|\alpha\rangle_t = E(t)|\alpha\rangle_t \tag{10}$$

where $E(t)$ is the energy and $|\alpha\rangle_t$ is the time-dependent basis in the energy representation. In general, we can use an expansion

$$|\Psi(t)\rangle = \sum_{\alpha} b_{\alpha}(t)|\alpha\rangle_t, \tag{11}$$

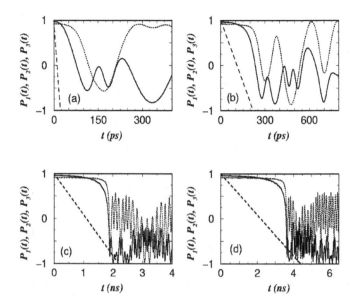

Fig. 5.3. Time evolution of the polarizations $P_1(t)$, $P_2(t)$, and $P_3(t)$ in a linear chain of three cells. The different plots correspond to different switch times of the driver cell: (a) $\tau = 10$ ps, (b) $\tau = 220$ ps, (c) $\tau = 2.2$ ps, (d) $\tau = 4.4$ ns. Reprinted with permission from C.-K. Wang, I. I. Yakimenko, I. V. Zozoulenko, and K.-F. Berggren, *J. Appl. Phys.* **84**, 2684 (1998). Copyright 1998, American Institute of Physics.

and then change the energy representation to the site representation given by Eq. (5) by means of the transformation,

$$c_{k,l}(t) = \sum_{\alpha} b_{\alpha}(t) c^{\alpha}_{k,l}(t), \tag{12}$$

where $c^{\alpha}_{k,l}(t)$ follows from the expansion

$$|\alpha\rangle_t = \sum_{k,l} c^{\alpha}_{k,l}(t)|k,l\rangle. \tag{13}$$

All we have done so far is to introduce an alternative representation that leaves the physics unchanged. To deal with the energy dissipation due to inelastic processes, we now introduce a set of phenomenological equations,

$$\hbar\frac{d(b_\alpha^* b_\alpha)}{dt} = -\gamma b_\alpha^* b_\alpha, \qquad \alpha = 2, \cdots, 16,$$

$$\hbar\frac{d(b_1^* b_1)}{dt} = \gamma(1 - b_1^* b_1), \qquad (14)$$

where the normalization is chosen as

$$\sum_{\alpha=1}^{16} b_\alpha^* b_\alpha = 1 \qquad (15)$$

and γ is a decay constant that is assumed to be equal for all excited states. The time-dependent equations for $b_\alpha(t)$ can then be written as

$$\hbar\frac{db_\alpha(t)}{dt} = -iE_\alpha(t)b_\alpha(t) - \frac{\gamma}{2}b_\alpha(t), \qquad \alpha = 2, \ldots, 16,$$

$$\hbar\frac{db_1(t)}{dt} = -iE_1(t)b_1(t) + A + iA, \qquad (16)$$

where $E_\alpha(t)$ is the αth eigenvalue at time t and

$$A = \frac{\gamma(1 - b_1^*(t)b_1(t))}{b_1^*(t) + b_1(t) + 2Im(b_1(t))}. \qquad (17)$$

The numerical solution of these equations can be obtained in the same fashion as for Eq. (6). That is, the polarization of the driver is assumed to be changing linearly with time from some prescribed value at $t = 0$ to another one at $t = \tau$, the switch time of the driver cell. The response of each separate cell is then calculated at time steps of Δt in the field of the surrounding cells. The time step has to be sufficiently small to ensure convergence. The relaxation time ($\tau_\gamma = \hbar/\gamma$) has been taken as $\tau_\gamma = 5.5$ ps, a reasonable value for phonon scattering in a semiconductor quantum dot.[7]

The results of our calculations show the improvement of the switching properties when effects of inelastic scattering are introduced explicitly. This is, of course, a positive aspect, since processes of this kind will always be present in a real device.

To mimic a somewhat more realistic switching mode for a QCA cell, we let the polarization in the driver cell develop in time in the following way. It is first changed from $P_1 = 0$ at $t = 0$ to $P_1 = -1$ with $\tau = 22$ ps and then held in this state for another 22 ps, after which it is driven further from $P_1 = -1$ to $P_1 = +1$ using $\tau = 44$ ps. Figs. 5.4(a) and (b) refer to the cases of two and three cells, respectively. The polarization of the driven cells now closely follows the imposed switching. When the driver reaches its saturation polarization (here -1 or $+1$), the other cells will consecutively

reach their saturation values as well, with only short delays. In this respect
the switching characteristics of our model QCA is quite satisfactory. It is
interesting to note that the saturation polarization of the second cell in the
case of two cells is $P_2^{sat} = \pm\ 0.92$, while for three cells $P_2^{sat} = \pm\ 0.97$, i.e.
somewhat higher than for two cells. The reason is that the second cell in
the chain with three cells couples with two neighboring cells, i.e. with both
the driver and the third cell, and therefore experiences a stronger external
force. This situation favors the consecutive transmission of the polarization
signal to the next cell in the chain. Hence the saturation value of the third
cell is still $P_3^{sat} = \pm\ 0.92$, i.e. the transfer of information is as good as for
the shorter chain. In general, the saturation polarization is related to the
property of the cell, namely, the tunneling coefficient, the Hubbard energy,
and the Coulomb energy. A detailed discussion is given in Ref. [3].

We have also studied the time evolution of the polarizations of longer
chains. Figure 5.5(a) shows, as an example, the case of a linear chain containing six cells. Here the relaxation time is taken as $\tau_\gamma\ =\ 3.5$ ps. The
distance between neighboring dots is taken as 34 nm and $\hbar\omega_x\ =\ \hbar\omega_y\ =$
18 meV, as above. The parameters are then calculated as $L \simeq 0.038$ meV,
$V_Q \simeq 1.7$ meV, and $E_Q \simeq 8.53$ meV. The polarization of the driver cell
is changed from $P_1 = 0$ at $t = 0$ to $P_1 = -1$ with $\tau = 18$ ps and then
from $P_1 = -1$ to $P_1 = +1$ with $\tau = 28$ ps. In this case the polarization of
the driven cells follows the polarization of the driver cell. The polarization
of the end cell is again somewhat less than those of the other cells because it experiences weaker external forces from only one nearest-neighbor
cell. This is a common property of cell switching in linear chains. We have
also investigated the straight chains with four and five cells and L-shaped
configurations, and found equally satisfactory switching behavior. For an
illustration, we show in Fig. 5.5(b) the time evolution of the polarization
for a symmetric L-shaped chain with five cells.

We have therefore shown that the QCA model with explicit incorpo-
ration of inelastic processes, unavoidable in real semiconductors, exhibits
excellent switching properties. As we will discuss below, imperfections will,
however, drastically modify this scenario in a negative way. The model,
described above, has been extended to the case of QCA in a thermal bath
with finite temperature.[8,9] To describe the dissipation mechanisms more
rigorously, in [8,9] the Markovian approach to the dynamical evolution of
the reduced density matrix was developed, where direct calculation of the
transition rates for the electron-phonon interaction is possible. As a result,
it has been found that the system response to the driving field is improved

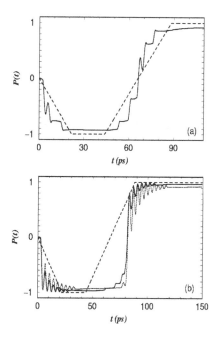

Fig. 5.4. Time-dependent behavior of the polarization in the different cells when energy dissipation is introduced explicitly. The decay time is $\tau_\gamma = 5.5$ ps. The polarization of the driver cell is varied linearly from 0 to -1 and from -1 to $+1$ (dashed line). Case (a) refers to a chain with two cells. The second cell (solid line) closely matches the changes in the driver cell. Case (b) corresponds to a chain with three cells and the same type of driver switching and decay time. Both the second cell (dashed line) and the third cell (dotted line) have polarizations aligning with that in the driver cell. Reprinted with permission from C.-K. Wang, I. I. Yakimenko, I. V. Zozoulenko, and K.-F. Berggren, *J. Appl. Phys.* **84**, 2684 (1998). Copyright 1998, American Institute of Physics.

at low temperature (and/or weak phonon coupling), before deteriorating as temperature and asymmetry increase.

5.5. Imperfections: Variable Coupling Strength, Defects, Stray Charges

In the previous sections we have considered the dynamics of regular arrays. In reality, however, semiconductor QCA devices are never free from fabrication imperfections. Therefore, the intriguing question arises about how

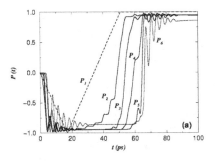

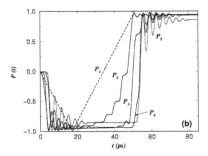

Fig. 5.5.   Time-dependent behavior of the polarization in the different cells when energy dissipation is introduced explicitly. Case (a) corresponds to a chain with six cells. Case (b) corresponds to a symmetric L-shaped chain consisting of five cells. For this calculation the distance between nearest-neighbor dots is taken as $a = 34$ nm. The parameters L, E_Q, and V_Q are thus calculated as 0.038 meV, 8.53 meV, and 1.70 meV, respectively. The decay time is $\tau_\gamma = 3.5$ ps. The polarization of the driver cell is changed linearly from 0 to -1 and from -1 to $+1$ (dashed line). Reprinted with permission from C.-K. Wang, I. I. Yakimenko, I. V. Zozoulenko, and K.-F. Berggren, *J. Appl. Phys.* **84**, 2684 (1998). Copyright 1998, American Institute of Physics.

such imperfections affect their proper operation. Furthermore, the study of imperfections is obviously linked with such problems as possible unidirectionality and metastability of QCA systems, which are widely discussed in the literature.[10] To solve these problems and in view of an operational model, one should therefore consider the dynamics directly, rather than focusing on steady-state calculations. We have studied the influence of imperfections stemming from:

(i) changes of intercellular distance;

(ii) changes of the height of the interdot barriers;

(iii) presence of stray charges.

Our studies have been based on using the model of QCA chain with dissipation outlined in Sec. 5.4.

5.5.1. *Variations of the intercellular distances*

To study the influence of the irregularities in the spatial displacement of the cells on the QCA dynamics let us consider the Hamiltonian (1) with

$$V_m \equiv V_m(t) = \sum_j \sum_{m'} V_Q \frac{(\rho^j_{m'}(t) - \rho)}{|\mathbf{R}_{j,m'} - \mathbf{R}_{i,m}|}, \tag{18}$$

where $j \neq i$. This expression defines the potential at site m in the i-th cell due to the charges in all the other cells j, ρ is a fixed positive charge with magnitude $(1/2)|e|$, which is introduced for providing electroneutrality of the cell as a whole (it is essential to ensure charge neutrality, if not, our simulations show that proper switching cannot be achieved).

In the case of a spacing error (defect) somewhere in the chain, V_m will be affected. This leads to a slight deviation of the Hamiltonian from its form for the regular chain, and the question arises about how much defects of this kind influence the temporal evolution of the polarization. We consider a cell which is placed relatively far away from the left driver cell, whose polarization, at some moment, is changed. Obviously, such a cell experiences a weaker force from its left neighbor than from the right one and, as a result, it can loose the ability to follow the polarization of the left cell, i.e. the chain may not switch into the new state as the polarization of the driver is changed. What is the limit of the deviation in the intercellular distances for which the chain still switches properly? This quantitative question can be answered by direct calculations of the temporal evolution for each specific case.

We have studied numerically the time-dependent behavior of the polarization in chains containing three and six cells, assuming the normal intercellular distance equal to 68 nm, and the decay time $\tau_\gamma = 3.5$ ps. The parameters involved in Eq. (1) have been evaluated using the simple two-dimensional parabolic confinement potential of Eq. (2) as $L = 0.038$ meV, $E_Q = 8.53$ meV and $V_Q = 1.7$ meV. We assume that the polarization of the driver cell is changed linearly from 0 to -1 and from -1 to +1 to mimic a real switching process. The time step is chosen small enough to ensure convergence. The results of the calculations for a chain containing six cells are

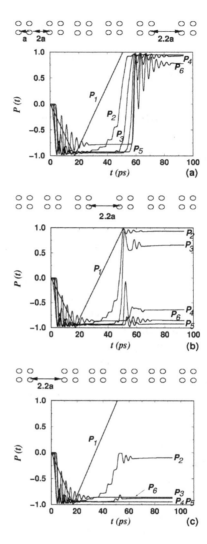

Fig. 5.6. Influence of the irregularities in the intercellular distance on the time-dependent behavior of the polarization for a chain consisting of six cells. The distance between nearest neighboring dots is $a = 34$ nm. The initial distance between dots in nearest neighboring cells is $d = 2a = 68$ nm. The parameters L, E_Q, and V_Q are 0.038 meV, 8.53 meV, and 1.70 meV, respectively. A spacing error of $+10\%$ ($d \rightarrow 1.1d$) is introduced between the fifth and sixth cells (a), the third and fourth cells (b), and the first (driver) and second cells (c). Reprinted with permission from I. I. Yakimenko, I. V. Zozoulenko, C.-K. Wang, and K.-F. Berggren, *J. Appl. Phys.* **85**, 6571 (1999). Copyright 1999, American Institute of Physics.

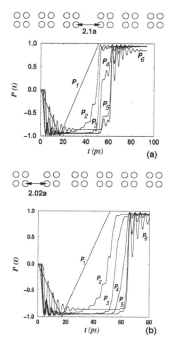

Fig. 5.7. Same as Fig. 5.6, but for a spacing error of +5% between the third and fourth cells (a) and of 1% between the driver and second cell (b). Reprinted with permission from I. I. Yakimenko, I. V. Zozoulenko, C.-K. Wang, and K.-F. Berggren, *J. Appl. Phys.* **85**, 6571 (1999). Copyright 1999, American Institute of Physics.

shown in Figs. 5.6 and 5.7. Figure 5.6(a) corresponds to the case in which the distance between the 5th and the 6th cells exceeds the regular intercellular distance by 10%, i.e. the irregular spacing is placed between the two cells at the end of the chain. In this case the cells maintain good switching properties in spite of the presence of a spacing error. However, if we put the defect in the middle of the chain, only the switching properties of the first and second cells remain satisfactory while, switching of the remaining cells completely fails (Fig. 5.6(b)). Finally, the situation is changed drastically if we put the spacing error at the beginning of the chain, i.e. between the driver cell and its nearest neighbor: in this case the the operation of the chain fails completely (Fig. 5.6(c)). At the same time, if the spacing error is scaled down to 5% in the middle of the chain (Fig. 5.7(a)) or to 1% at the beginning of the chain (Fig. 5.7(b)), the chain switches properly. We only observe some delay in the switching of the cells situated to the right of the defect. These values represent the limit of permissible errors for the intercellular distances.

Therefore, we emphasize the great sensitivity of switching properties of the long chains to the presence of spacing errors, especially when errors occur near the end of such chains. For comparison, we have also studied the influence of a change of intercellular distance on the switching of a chain with three cells. In this case the influence of the error is not so critical: even with a spacing error of 35% occuring between the driver and first cell, switching still occurs (Fig. 5.8). This percentage defines the sharp boundary between functionality and non-functionality of the array.

To understand the increased sensitivity of longer chains, it is instructive to inspect the level spacing for steady states. We then find that shorter chains are simply more robust because of their larger level spacings (see also Ref. 11). The magnitude of the typical energy splitting is 10^{-5} eV, which leads to the higher sensitivity of longer chains to irregularities of the intercellular distance (which cause energy shifts of the same order of magnitude as the splitting, and hence mixing of states).

This effect becomes more pronounced when the polarization of the driver cell is small. This situation always occurs in the dynamic case, when the polarization of the driver cell in varied continuously, and thus passes through

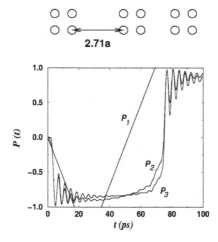

Fig. 5.8.   Same as Fig. 5.6, but for a chain of three cells, with a spacing 35% larger between the driver cell and the first driven cell. Reprinted with permission from I. I. Yakimenko, I. V. Zozoulenko, C.-K. Wang, and K.-F. Berggren, J. Appl. Phys. **85**, 6571 (1999). Copyright 1999, American Institute of Physics.

$P_1 = 0$. Therefore time-dependent simulations show greater sensitivity to irregularities in an array than time-independent ones do. In order to avoid the point where the polarization of the driver is near zero, Lent *et al.*[5] have proposed another operational mode, the so called adiabatic switching, in which the polarization of the driver cell changes simultaneously with the height of interdot barriers.

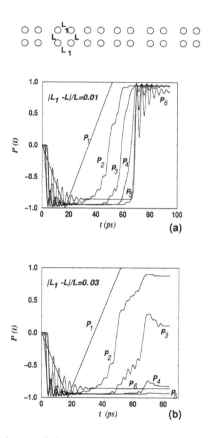

Fig. 5.9.   Temporal evolution of the polarization in the six-cell chain in the presence of defects of the interdot barriers in the second cell: (a) $|L_1 - L|/L = 0.01$; (b) $|L_1 - L|/L = 0.03$. Reprinted with permission from I. I. Yakimenko, I. V. Zozoulenko, C.-K. Wang, and K.-F. Berggren, *J. Appl. Phys.* **85**, 6571 (1999). Copyright 1999, American Institute of Physics.

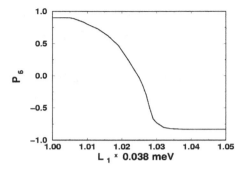

Fig. 5.10. Dependence of the polarization of the output cell on the tunneling parameter L_1 in the second cell for the chain with six cells. Reprinted with permission from I. I. Yakimenko, I. V. Zozoulenko, C.-K. Wang, and K.-F. Berggren, *J. Appl. Phys.* **85**, 6571 (1999). Copyright 1999, American Institute of Physics.

5.5.2. *Defects in interdot barriers*

The next case of interest is a regular chain with the interdot barriers changed in some cell in such a way that $L_{1,2} = L_{3,4} = L$ and $L_{2,3} = L_{4,1} = L_1 > L$, where we assume $L = 0.038$ meV, as in the calculations in the preceding section.

As the tunneling energies are increased, the two-electron wave function becomes less localized in the antipodal sites in the cell, and the polarization of this cell becomes less than ± 1, which has an influence on the following cells. Thus one can expect that even a completely polarized driver fails to polarize the chain. In other words, a cell with such a defect turns out to be a severe obstacle for propagation of the polarization along the chain.

This expectation is confirmed by Figs. 5.9(a) and (b), showing the temporal evolution for the six-cell chain with the defect in the second cell consisting in two fixed values of the tunneling constant: $|L_1 - L|/L = 0.01$ (a) and $|L_1 - L|/L = 0.03$ (b). The dependence of the output polarization on the tunneling parameter L_1 is shown in Fig. 5.10. Note that, if the value L_1 is about 0.0399 meV, the polarization of the output cell falls quickly to zero. The transition to a non-switching regime occurs near $L_1 = 0.04$ meV, which corresponds approximately to the Coulomb-induced splitting between the energy levels of oppositely polarized states (which is about 0.04 meV for our system). Thus, if the interdot barriers in the cell adjacent to the driver are changed only by 5%, the switching properties of a chain as a whole fail. This is in agreement with the results for the steady-state calculations obtained in Ref. 12.

The situation is not so crucial for the defects occurring in the middle (Fig. 5.11(a)) or at the end (Fig. 5.11(b)) of the chain. The corresponding threshold values of the tunneling parameters L_1 are 2.5L (Fig. 11 (a)) and 3L(Fig. 11 (b)). In the case of the three-cell chain with the defect in the second cell we found that the switching properties are not lost for a value of the tunneling constant up to 2L. All these quantitative results provide some information about the extent to which fluctuations in the interdot barrier height influence the switching properties of QCA circuits.

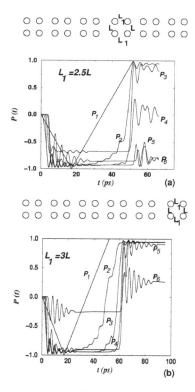

Fig. 5.11. Temporal evolution of the polarization for the six-cell chain in the presence of defects of the interdot barriers in the fourth cell (a) and in the sixth cell (b). Reprinted with permission from I. I. Yakimenko, I. V. Zozoulenko, C.-K. Wang, and K.-F. Berggren, *J. Appl. Phys.* **85**, 6571 (1999). Copyright 1999, American Institute of Physics.

5.5.3. *Effect of stray charges*

Quite a common type of imperfection in real QCA arrays is the presence of stray charges. Here we consider the effect of such charges when located on the same plane as the chain. If the charge is close to the chain, it introduces strong asymmetry into the energy levels, which may completely erase the effect of Coulomb interaction coming from the neighboring cells. For instance, inserting a point impurity with charge e at a distance D near the middle of a chain with six cells, we have obtained incorrect switching when the distance between the electron and the chain is less than 400 nm. A three-cell array shows a switching failure when the stray charge appears at a distance less than 140 nm. It appears that the transition between the two regions, in which the chain either works or fails, is rather sharp – a few nanometers only. This sharpness has also been found by Lent *et al.*,[13]

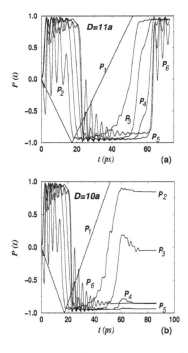

Fig. 5.12.   Influence of a stray charge on the polarization in a chain with six cells: (a) $D = 11a$; (b) $D = 10a$. Reprinted with permission from I. I. Yakimenko, I. V. Zozoulenko, C.-K. Wang, and K.-F. Berggren, *J. Appl. Phys.* **85**, 6571 (1999). Copyright 1999, American Institute of Physics.

who obtained it within the steady-state approximation. However, critical distances depend sensitively on the parameters of the dots and on the number of cells. To illustrate this, we reproduce in Fig. 5.12 an example of the temporal evolution of the polarization in the case of a six-cell array. Even small changes in the distance between the chain and the perturbing stray charge do induce strong changes in the polarization of the cells. Therefore, the results of our numerical study of the dynamical response of imperfect QCA chains to the polarization change of a driver cell clearly demonstrate that the operation of long chains of QCA cells is very sensitive to the presence of irregularities in the spatial distribution of cells, of local variations of the interdot barriers, and of stray charges. For example, spacing errors (especially at the beginning of the chain) or a stray charge located at up to five intercellular distances can completely destroy switching of a chain with six cells. To implement semiconductor based QCA circuits in practice, one often thinks of close-packed arrays of cells with many chains operating at the same time. The electrons in one chain will then act as perturbing stray charges for adjacent chains. In view of the extreme sensitivity of a single chain, the entire array easily becomes very volatile under such circumstances.

References

1. C.-K. Wang, I. I. Yakimenko, I. V. Zozoulenko, and K.-F. Berggren, *J. Appl. Phys.* **84**, 2684 (1998).
2. I. I. Yakimenko, I. V. Zozoulenko, C.-K. Wang, and K.-F. Berggren, *J. Appl. Phys.* **85**, 6571 (1999).
3. C. S. Lent, P. D. Tougaw, and W. Porod, *J. Appl. Phys.* **74**, 3558 (1993); C. S. Lent and P. D. Tougaw, *J. Appl. Phys.* **74**, 6227 (1993); P. D. Tougaw and C. S. Lent, *J. Appl. Phys.* **75**, 1818 (1994).
4. P. D. Tougaw and C. S. Lent, *J. Appl. Phys.* **80**, 4722 (1996).
5. C. S. Lent and P. D. Tougaw, *Proc. IEEE* **85**, 541 (1997).
6. G. H. Bernstein, G. Bazan, M. Chen, C. S. Lent, J. L. Merz, A. O. Orlov, W. Porod, G. L. Snider, and P. D. Tougaw, *Superlattices and Microstructures* **20**, 447 (1996).
7. T. Schmidt *et al.*, *Phys. Rev. Lett.* **78**, 1544 (1997); T. Inoshita and H. Sakaki, *Phys. Rev. B* **56**, R4355 (1997).
8. F. Ramirez, E. Cota, S. E. Ulloa. *Phys. Rev. B* **62**, 1912, (2000).
9. F. Rojas, E. Cota, S. E. Ulloa. *Phys. Rev. B* **66**, 235305, (2002).
10. S. Bandyopadhyay, B. Das, A.E. Miller, *Nanotechnology*, **5**, 113 (1994); R. Landauer, *Phil. Trans. Roy. Soc. Lond.* **A353**, 367 (1995).
11. M. Governale, M. Macucci, G. Iannaccone, C. Ungarelli, J. Martorell, *J. Appl. Phys.* **85**, 2962 (1998).

12. M. Girlanda, M. Governale, M. Macucci, G. Iannaccone. *J. Appl. Phys.* **75**, 3198, (1999).
13. P. D. Tougaw and C. S. Lent, *Jpn. J. Appl. Phys.* **34**, 4373 (1995).

CHAPTER 6

Time-Dependent Analysis of QCA Circuits with the Monte Carlo Method

Luca Bonci, Mario Gattobigio[a], Giuseppe Iannaccone
and Massimo Macucci

Dipartimento di Ingegneria dell'Informazione
Università di Pisa
Via Caruso 16, I-56122 Pisa, Italy

6.1. Introduction

As previously discussed, the so called "ground state computation" principle is at the base of the simplest implementation of QCA logic.[1] This may lead to serious problems, as we shall see in the following, when the dynamical behavior of large circuits is considered. Metastable states, corresponding to the wrong logical output, i.e. inconsistent with the inputs, do indeed exist and may prevent the system from reaching the actual ground state for a time that is long compared with the desirable completion time of the computation.[2]

Craig Lent and coworkers[3] have proposed a scheme to circumvent this problem, which is related to the parametron concept introduced by Korotkov and Likharev;[4] it is based on driving the system evolution via a multi-phase clock, making sure that when a cell is acting as a driver cell it is in a "locked" state, so that its polarization cannot be perturbed, while the driven cell goes through an "adiabatic" switching process, evolving through quasi-equilibrium states, which guarantees the absence of metastable states.

In the following, we study a specific proposed implementation of this method[5] by performing analytical and numerical calculations with the goal of assessing its practical feasibility. In particular, we consider a realistic cell

---

[a]Current address: Scuola Normale Superiore, Piazza dei Cavalieri, I-56126 Pisa, Italy

geometry and determine the region of the parameter space, depending on material choice, that allows reaching a clock speed acceptable for real-life applications. We then move on to the simulation of a few circuits based on such a cell and establish the maximum achievable clock frequency, finding that it is orders of magnitude smaller than the reciprocal of the RC time constant of the system. In addition, the time-dependent performance of adiabatic circuits is compared with that of nonclocked circuits.

6.2. Six-Dot QCA Cell

Clocking a QCA cell basically consists in a two step process. In the first step the cell evolves quasi-adiabatically toward the final ground state, while the potential barriers separating the dots are gradually raised. Then the cell is "frozen" (by raising the barriers to their maximum height), so that its polarization state cannot change and can be used to drive the nearby cells.

A method to modulate the inter-dot barrier in a QCA implementation based on metal dots has been proposed in Ref. 5, where a metal dot QCA is considered. This implementation allows for replacing the barrier with another dot whose potential can be varied by means of an external bias. The complete cell, shown in Fig. 6.1, is made up of six dots: four of them

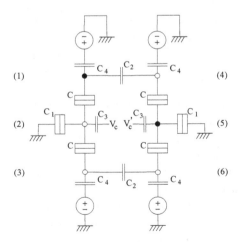

Fig. 6.1.   Six-dot cell. Each semi-cell consists of three dots connected by tunneling junctions and coupled to external voltage sources via capacitors. Reprinted with permission from L. Bonci, G. Iannaccone, and M. Macucci, *J. Appl. Phys.* **89**, 6435 (2001). Copyright 2001, American Institute of Physics.

(1,3,4, and 6) encode the logical state as in the ordinary QCA scheme, while the other two (2 and 5) are required for modulating the inter-dot barriers.

The operation of this new kind of QCA cell can be understood focusing on half a cell, e.g. the one containing dots 1, 2 and 3 (left semicell). The three integers $[n_1, n_2, n_3]$ indicating the occupation number of the three dots describe the cell status that, in operating conditions, namely when only an excess electron is contained in the semicell, can assume the following three configurations: $[1, 0, 0]$, $[0, 1, 0]$ and $[0, 0, 1]$.

The configuration energy depends on the value of the control voltage V_c. If this parameter is kept at large negative values, the $[0, 1, 0]$ configuration energy can be made much higher than the energies of the other two configurations, creating a potential barrier that prevents the electron from moving from dot 1 to dot 3, and thus *locking* the semicell. An opposite condition holds when V_c is largely positive and thus when the central dot represents a potential hole for the excess electron. In this case the electron stays in dot 2 and the semicell is said to be in a *null* state, which means that there is no excess electron with a specific influence on the polarization of the nearby cell or on the other semicell.

Finally, we have the *active* condition, that correspond to a control voltage making the $[0, 1, 0]$ configuration energy comparable to the energy of the other two configurations. In the active condition, the excess electron can move across the semicell as a function of the external (of the nearby semicell and neighboring cells) configurations.

The operation of the semicell is the following (see Fig. 6.2), as discussed by Lent *et al.*[5] Starting from the null condition, in which the semicell does not influence the neighboring dots, the control voltage is lowered so that the semicell enters the active region and can easily reach a local ground state, which depends on the charge configuration in nearby cells. Once the semicell has relaxed to the ground state, V_c is lowered again, to reach the locked state. At this point the electronic configuration of the semicell can be used (without any risk of being modified by the interaction) to act upon the polarization state of nearby cells. The same procedure is followed if we consider the full cell as a single element, driving with the same voltage the two control electrodes.

The data flow in a QCA system can thus be steered by applying an appropriate clock sequence and, at the same time, it is possible to avoid the problem associated with the presence of metastable states. This can be achieved by modulating the barriers with voltages that vary slowly enough to allow the cell to continuously be in the instantaneous ground state (adi-

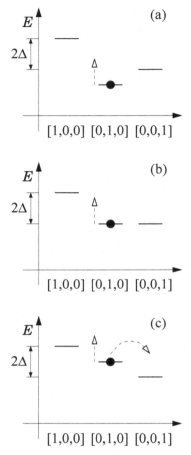

Fig. 6.2.   Representation of the energy for the different configurations and of the tunneling path for three consecutive time steps. The $[0, 1, 0]$ energy depends on the the value of the control voltage V_c'. Reprinted with permission from L. Bonci, G. Iannaccone, and M. Macucci, *J. Appl. Phys.* **89**, 6435 (2001). Copyright 2001, American Institute of Physics.

abatic switching). The speed of signal propagation and processing will be limited by the switching time of single cells, which has to satisfy the adiabatic requirement and also be long enough compared to the average time that elapses between tunneling events. If the latter condition were not satisfied, the cell would not be able to settle in the ground state during the active period of the clock, because tunneling is a statistical process, and it will occur with probability close to 1 only if a time much larger that the inverse of the tunneling rate has elapsed.

To assess the maximum clock speed allowed for the correct operation of a QCA device, we thus first need to evaluate the tunneling rate in the active region. In the following such a calculation will be performed, assuming as reference configuration the one with the [0,1,0] energy intermediate between those of the [1,0,0] and [0,0,1] configurations (see Fig. 6.2(c)). Our model includes a simplified cell geometry, to compute the dependence of the tunneling rate on cell size and material parameters.

6.2.1. *Transition rates for a semi-cell*

In order to derive a quasi-analytical expression for the tunneling rate, we have made several simplifying assumptions, but we have always intentionally made approximations that lead to an overestimate of the rate. In such a way, we obtain optimistic predictions for the maximum clock speed. Energy quantization in the dots has been neglected, considering that the prevalent energy is of electrostatic nature. Moreover, we considered the inter-dot barrier thickness much smaller than the dot size, in order to justify a one-dimensional model for the tunneling process, i.e. to be able to consider only the direction normal to the junction, as shown in Fig. 6.3. This last approximation also leads to an overestimation of the tunneling rate, since it implies perfect overlap between transverse states.

Referring to Fig. 6.3, regions 1 and 2 represent the central and side dots, respectively, and the barrier is represented by the insulating medium separating the two dots. Important parameters in the definition of the tunneling rate are the barrier size, more precisely its thickness a and height

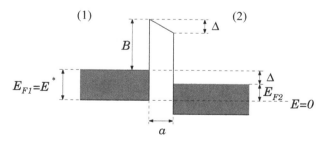

Fig. 6.3. Conduction band profile used to estimate the tunneling rate between the central dot and one of the side dots. We show the typical biasing condition corresponding to Fig. 6.2(c).Reprinted with permission from L. Bonci, G. Iannaccone, and M. Macucci, *J. Appl. Phys.* **89**, 6435 (2001). Copyright 2001, American Institute of Physics.

B, the effective electron mass, both in the dot, m, and inside the barrier m_B, the energy offset Δ, and the distance between the Fermi level and the bottom of the conduction band in the dots, which we have indicated as E^*. By assuming the energy reference at the bottom of the conduction band of dot 1, we can define $E_{F1} \equiv E^*$ and $E_{F2} \equiv E^* - \Delta$, where E_{F1} and E_{F2} are the quasi Fermi energies of the two dots (see Fig. 6.3). A further parameter we need is the dielectric permittivity ϵ of the insulator.

Let us consider an electron with transverse wave vector $k_\parallel$ and energy E_z along the direction normal to the junction (the z-axis). By assuming conservation of spin, total energy, and transverse wave vector during tunneling, the current density across the barrier can be written as

$$ J = \frac{2e}{\hbar} \int dE_z \int \frac{d^2 k_\parallel}{(2\pi)^3} 2T(E_z) \left[F(E, E_{F1}) - F(E, E_{F2}) \right] , \qquad (1) $$

where $F(E, E_{Fi})$ is the occupation probability for a state of energy E in dot i, the factor of 2 takes into account spin degeneracy, e is the electron charge, and $\hbar$ is the reduced Planck constant. The barrier transmission probability is denoted as $T(E_z)$, which in our case does not depend on the transverse wave vector. The integral is over all $\vec{k}_\parallel$ and $E_z > 0$.

In the low temperature limit, the occupation probability can be described by a step function:

$$ F(E, E_{F1}) - F(E, E_{F2}) = \Theta(E - E_{F2})\Theta(E_{F1} - E) $$
$$ = \Theta(E - E_{F1} + \Delta) \times \Theta(E_{F1} - E) . \qquad (2) $$

To keep calculations simple, we assume parabolic bands, thus with $E = E_z + \hbar^2 k_\parallel^2/(2m)$, and we consider for E^* an effective value yielding a reliable estimate of the density of states at the Fermi energy. With a change of variables, we get $\eta^2 \equiv \hbar^2/(2m)(k_x^2 + k_y^2)$ and $\phi \equiv \text{arctg}(k_y/k_x)$, so that Eq. (1) becomes

$$ J = \frac{me}{4\pi^3 \hbar^3} \int dE_z T(E_z) \int_0^{2\pi} d\phi \qquad (3) $$
$$ \times \int d\eta\, \eta\, \Theta(E^* - E_z - \eta^2)\Theta(E_z - E^* + \eta^2 + \Delta) . $$

If $E_z < E^* - \Delta$, the integration domain consists of the region comprised between two circles of radius $\eta_1 = \sqrt{E^* - E_z - \Delta}$ and $\eta_2 = \sqrt{E^* - E_z}$, while, for $E^* - \Delta < E_z < E^*$, the domain is the circle of radius $\eta = E^* - E_z$.

With a polar integration, we get:

$$J = \frac{me}{2\pi^2\hbar^3} \left[\int_0^{E^*-\Delta} dE_z T(E_z)\Delta - \int_{E^*-\Delta}^{E^*} dE_z T(E_z)(E^* - E_z) \right]. \quad (4)$$

In order to write an explicit form for the transmission coefficient, we have limited our analysis to the case of a barrier much higher than the electron energy, thus obtaining

$$T(E_z) \approx \frac{E_z}{B} \exp\left[-\frac{2a}{\hbar}\sqrt{2m_B(B_{\text{eff}} - E_z)} \right], \quad (5)$$

where B_{eff} is the effective barrier height for an electron with longitudinal energy E_z. If the bias does not cause an appreciable deformation of the barrier, we can simply consider its average height $B_{\text{eff}} = B + E^* - \Delta/2$. We evaluate the integrals in Eq. (4) numerically.

If electrons tunnel between dots over an area S, the transition rate is given by

$$\Gamma = JS/e. \quad (6)$$

The current density and the rate Γ will have a linear dependence on Δ, if $\Delta \ll E^*$. If instead Δ becomes close to E^*, a quasi-saturation of the current occurs (for $\Delta = E^*$ the first integral in Eq. (4) and the dependence on Δ disappear). For larger bias values, the current density still depends on

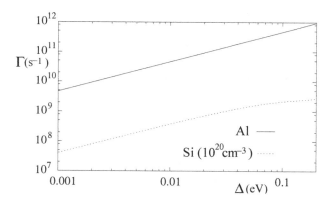

Fig. 6.4. Tunneling rate as a function of the energy offset Δ. The two results correspond to an Al dot (solid line) and to an n-doped ($n = 10^{20}$ cm^{-3}) Si dot (dashed line). In both cases we choose SiO$_2$ as dielectric and the value of the parameters are: $a = 2$ nm and $S = 100$ nm^2. Reprinted with permission from L. Bonci, G. Iannaccone, and M. Macucci, *J. Appl. Phys.* **89**, 6435 (2001). Copyright 2001, American Institute of Physics.

Δ, but only through the dependence on Δ of T, the transmission coefficient. In Fig. 6.4 we represent Γ vs. Δ for aluminum dots and Si dots with high doping, considering a SiO_2 barrier $a = 2$ nm thick and with a cross-section of $S = 100$ nm^2. The two cases differ mainly for the value of E^*. For the case of Al dots, we have used the measured free electron density, and, following Ref. 6 we have obtained $E^* = 11.7$ eV, while for case of silicon dots the parabolic band approximation yields $E^* \approx 0.15$ eV. In the latter case saturation of Γ is reached for values of Δ of the same order as E^*. The zero temperature tunnel resistance is finally given by $R_T = \Delta/(e^2\Gamma)$ for $\Delta \ll E^*$.[7]

6.3. Analysis of the Parameter Space

6.3.1. *Tunneling rate*

As shown in the previous section, the tunneling rate strongly depends on material and geometric parameters. We want to determine the regions in the parameter space in which it is possible to achieve a tunneling rate large enough to allow a clock frequency useful for practical purposes. For safe operation we need to assume a tunneling rate significantly larger than the switching frequency: we require that the tunneling rate be at least a hundred times larger than the clock rate. Therefore, in order to obtain a 10 MHz clock we need to look for a minimum transition rate $\Gamma_{min} = 10^9$ s^{-1}.

The rate described by Eq. (6) depends on several parameters, thus, before directing our attention to the the particular cell of Fig. 6.1, we need to devise a strategy for the targeted exploration of the multidimensional parameter space, which is otherwise too large for an exhaustive exploration. As a first step, let us take into consideration different material combinations and barrier characteristics. At each temperature T we must set the value of the energy imbalance Δ large enough to allow proper operation. A choice yielding a minimal protection against the adverse effect of thermal fluctuations is $\Delta = 10k_BT$, although this is not a trivial condition to be satisfied, as we will discuss below.

The barrier is described by three parameters: the height B, the electron effective mass inside the barrier m_B, which both depend on material choice, and the thickness a. In Figs. 6.5(a-d) we choose $a = 2$ nm, and plot the curves corresponding to $\Gamma = \Gamma_{min} = 10^9$ s^{-1}, given by (4) and (5). We made four different choices for the dot material and assumed two values of the operating temperature: $T = 4.2$ K (solid lines) and $T = 77$ K (dashed lines). In addition, two typical junction surface values are considered for

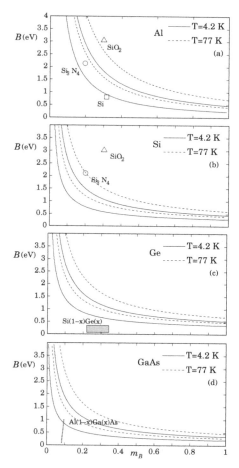

Fig. 6.5.   Barrier height B as a function of the effective mass m_B (expressed in terms of the electron mass) for the condition in which the tunneling rate $\Gamma = 10^9$ s^{-1} for dots in aluminum (a), silicon (b), germanium (c) and gallium arsenide (d). Results for $T = 4.2$ K (solid lines) and $T = 77$ K (dashed lines) are reported. The two curves for each temperature are computed for a junction surface S equal to $S = 10$ nm^2 (lower curve) and $S = 10^4$ nm^2 (upper curve). The barrier thickness is assumed to be $a = 2$ nm. We consider typical barrier and dot materials. The wide dashed region in (c) approximately indicates the range of SiGe parameters for varying strain and Ge concentration. The dashed thick line in (d) indicates the range of values for an Al$_{1-x}$Ga$_x$As alloy. Reprinted with permission from L. Bonci, G. Iannaccone, and M. Macucci, *J. Appl. Phys.* **89**, 6435 (2001). Copyright 2001, American Institute of Physics.

each temperature: $S = 10$ nm^2 and 10^4 nm^2, which correspond to the lower and upper curve, respectively. The acceptable region coincides with the area below each curve, where the transition rate is larger than 10^9 s^{-1}. Symbols, corresponding to B and m_B for typical barrier materials, have been included.

Semiconductors are considered to be in a degenerate condition, which means n$^+$-doped Si and Ge with $N_D = 10^{20}$ cm^{-3} and GaAs with $N_D = 10^{19}$ cm^{-3}. Degenerate semiconductors are characterized by an extended sub-band, formed by donor levels, which almost overlaps the conduction band. This implies an effective thinning of the energy gap and complete donor ionization, so that $n = N_D$ even at zero temperature.

For the sake of comparison, the same data have also been calculated with an extremely optimistic barrier thickness ($a = 1$ nm), see Figs. 6.6(a-d). Fabrication of such thin barriers is still beyond current technological capabilities, because, while SiO$_2$ layers thinner than 1 nm have been successfully fabricated, a process to build a 1 nm-thick layer in the lateral direction is not yet available.

Before discussing these results we need to clarify that the increase of the tunneling rate with temperature is simply the result of the previously mentioned constraints, i.e. we are changing Δ with temperature, in order to keep $\Delta = 10k_BT$. The increase of Δ also involves an increase of Γ. Satisfying this requirement poses constraints on cell geometry, which we will discuss in the following section.

As expected, a decrease in the barrier thickness a leads to a strong increase of the tunneling rate. If $a = 1$ nm and we use Al dots, a large enough rate can be achieved for virtually any combination of insulator, temperature and junction area (see Fig. 6.6(a)). However, this is substantially a theoretical result, since fabrication of structures with such a fine spatial resolution are not within the reach of medium term technological capabilities.

A somewhat more realistic structure is presented in Fig. 6.5, although these parameter values are not yet within the reach of fabrication technology. An overall conclusion is that the low number of carriers in semiconductor dots, even in the presence of very high doping, does limit the achievable tunneling rate. This is somewhat compensated by the tunneling barrier height B, which can be small and, within certain limits, can be adjusted, in the case, for example, of Al$_x$Ga$_{1-x}$As-GaAs or Si$_x$Ge$_{1-x}$-Ge.

Metal islands have the advantage of containing a very large number of carriers available for tunneling, but the tunneling rate is limited by the height of the barrier and the value of the effective mass in the barrier. A lower barrier height and a smaller effective mass in the barrier are possible with semiconductors, but there are few carriers available in each dot. Over-

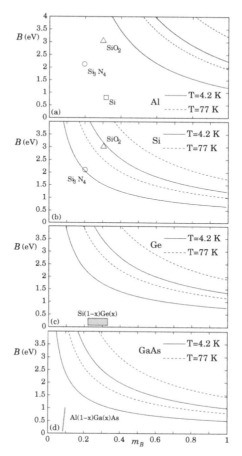

Fig. 6.6. Barrier height B as a function of the effective mass m_B (expressed in terms of the electron mass) for the condition in which the tunneling rate $\Gamma = 10^9$ s^{-1} for dots in aluminum (a), silicon (b), germanium (c) and gallium arsenide (d). Results for $T = 4.2$ K (solid lines) and $T = 77$ K (dashed lines) are reported. The two curves for each temperature are computed for a junction surface S equal to $S = 10$ nm^2 (lower curve) and $S = 10^4$ nm^2 (upper curve). The barrier thickness is assumed to be $a = 1$ nm. We consider typical barrier and dot materials. The wide dashed region in (c) approximately indicates the range of SiGe parameters for varying strain and Ge concentration. The dashed thick line in (d) indicates the range of values for an Al$_{1-x}$Ga$_x$As alloy. Reprinted with permission from L. Bonci, G. Iannaccone, and M. Macucci, *J. Appl. Phys.* **89**, 6435 (2001). Copyright 2001, American Institute of Physics.

all, the solution with metal islands seems to have better perspectives, also considering the possibility of using intrinsic semiconductors as dielectrics.

The wide range of small gap materials thus available, with low barrier height and low effective mass, would open up several degrees of freedom to play with in device optimization.

6.3.2. *Calculation of the energy imbalance*

For the cell we are considering, made up of six capacitively coupled metallic dots, a classical description is possible in terms of the values of five capacitors. The energy in a cell can be written as:

$$E = \frac{1}{2} \left(\mathbf{q} \; \mathbf{q}' \right) \mathbf{C}^{-1} \begin{pmatrix} \mathbf{q} \\ \mathbf{q}' \end{pmatrix} - \mathbf{v} \cdot \mathbf{q}' , \tag{7}$$

where $\mathbf{C}^{-1}$ is the inverse of the capacitance matrix, $\mathbf{v}$ is the vector of the voltages applied to the leads, while $\mathbf{q}$, $\mathbf{q}'$ are the vectors of the charge on the dots and on the leads, respectively. With Eq. (7) we can compute the energy difference 2Δ between two configurations, differing for the position of one electron (see Fig. 6.2). Furthermore, cell symmetry warrants that Δ is a function of only three capacitances: C, C_2 and C_4, as can be deduced from Fig. 6.1.

The region in which Δ exceeds a given value can be represented in a three-dimensional space, as shown in Fig. 6.7 where two possible values of Δ are considered: $\Delta = 3.6 \cdot 10^{-3}$ eV (upper surface) and $\Delta = 6.7 \cdot 10^{-2}$ eV

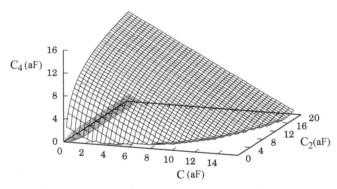

Fig. 6.7. The regions under the wire-mesh surfaces represent the portions of the parameter space $\{C, C_2, C_4\}$ in which $\Delta > 3.6 \cdot 10^{-3}$ eV (upper surface) and $\Delta > 6.7 \cdot 10^{-2}$ eV (lower surface). The two surfaces have been computed for an operating temperature of 4.2 K and 77 K, respectively. Reprinted with permission from L. Bonci, G. Iannaccone, and M. Macucci, *J. Appl. Phys.* **89**, 6435 (2001). Copyright 2001, American Institute of Physics.

(lower surface). These values have been chosen so that $\Delta = 10k_BT$ for $T = 4.2$ K and $T = 77$ K. Notice that the energy imbalance increases with C_2 while it decreases with C and C_4. As expected, see Fig. 6.7, the acceptable region is significantly reduced as Δ increases: for operation at 4.2 K capacitances of the order of a few attofarads are needed, whereas operation at 77 K is possible with capacitances around 1 aF. Although these values are in principle achievable, we must also consider the associated requirement of a very small barrier thickness, which may pose formidable technological challenges.

Other problems would derive from the choice of a large value of C_2, implying a large structure or a small intracell distance: this would finally lead to electrons tunneling between different cells, which is not acceptable for correct operation.

From the discussion so far, it is apparent that the same parameters affect at the same time the transition rates and the energy imbalance. In the following we focus on a well defined cell model, trying to verify the possibility of concurrently satisfying the different constraints.

The cell of Fig. 6.1 can be implemented with a simple geometry, with six parallelepiped dots located on a rectangular mesh, as shown in Fig. 6.8. The relevant parameters are the dot dimensions (l_1, l_2 and h), the distance between dots in a cell b, and the tunneling barrier width a. Other parameters that depend on material choice are the dielectric constant of the insulating medium, the tunneling barrier height B and the effective electron mass m_B in the barrier. Finally, we need to consider the values of E^* and m, which

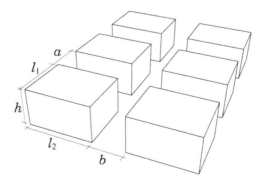

Fig. 6.8. Sketch of the considered cell geometry. Reprinted with permission from L. Bonci, G. Iannaccone, and M. Macucci, *J. Appl. Phys.* **89**, 6435 (2001). Copyright 2001, American Institute of Physics.

are determined by the choice for the dot material. We can assume, for the distances between dots, a lower limit of the order of a few nanometers, dictated by technological limitations. The other parameters may vary within a relatively wide range, owing to the various material combinations that can be used.

If we use, as a first order approximation, the parallel plate capacitor formula for the evaluation of the interdot capacitances, we can compute the electron tunneling rate making use of the results of Section 6.2.1 and calculating the electrostatic energy imbalance via Eq. (7).

We obtained results for a metal-insulator cell, made up of Al dots with three possible insulator choices (SiO_2, Si_3N_4, and Si) and two possible values of the barrier thickness a: 2 nm and 3 nm. We chose the following values for the other parameters: $b = 5$ nm and $l_1 = l_2 = h = 10$ nm. The quantities B, m_B and ϵ_r depend on the insulator, as shown in Table 6.1. The computed rate, energy imbalance Δ, and maximum operating temperature $T_{max} \equiv 10k_B/\Delta$ are shown in Tables 6.2, 6.3. The most critical parameters are the barrier thickness a, the barrier height B and the mass m_B,

Table 6.1.   Material parameters

	SiO_2	Si_3N_4	Si
B	3.05 eV	2.1 eV	0.8 eV
ϵ_r	3.9	6.7	11.1
m_B	0.3	0.2	0.33

Table 6.2.   Tunneling rate Γ and maximum operating temperature T_{max} for different insulator choices, for $a = 3$ nm.

	SiO_2	Si_3N_4	Si
Γ	$18.9\,s^{-1}$	$4.7 \times 10^5\,s^{-1}$	$1.1 \times 10^7\,s^{-1}$
Δ	1.5×10^{-2} eV	1.2×10^{-2} eV	8.6×10^{-3} eV
T_{max}	16.9 K	13.8 K	9.9 K

Table 6.3.   Tunneling rate Γ and maximum operating temperature T_{max} for different insulator choices, for $a = 2$ nm.

	SiO_2	Si_3N_4	Si
Γ	$4.5 \times 10^5\,s^{-1}$	$3.5 \times 10^8\,s^{-1}$	$2.1 \times 10^9\,s^{-1}$
Δ	9.8×10^{-3} eV	7.6×10^{-3} eV	5.1×10^{-3} eV
T_{max}	11.4 K	8.8 K	5.9 K

because the rate depends on them exponentially. For example, by simply reducing a down to 2 nm, we obtain for the previous example (with Si as an insulator) $\Gamma = 2 \times 10^9$ s^{-1}: an acceptable rate. While the barrier thickness can be reduced with better fabrication processes, m_B and B are intrinsic properties of the chosen material.

A further study can be undertaken in the $\{a, b, L\}$ parameter space, considering cubic dots with edge length L. The plot in Fig. 6.9 represents the surface that corresponds, in this space, to $\Gamma = \Gamma_{min} = 10^9$ s^{-1}. Below the surface, we obtain a tunneling rate larger than this limit, and thus we achieve the possibility of faster operation. The tunneling rate depends strongly on the distance a between dots, which must be very small, in order to achieve a reasonable tunneling rate. Curves for a few values of the maximum operating temperature T_{max} are plotted on the surface in Fig. 6.9.

The operating temperature is a relevant constraint: in order to obtain an energy imbalance Δ allowing operation at a few kelvins, extremely small dots (small L) and small inter-cell distances b are needed.

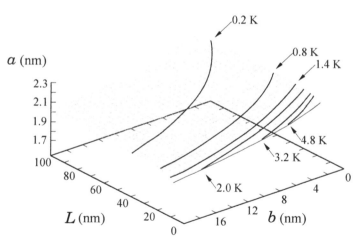

Fig. 6.9. The surface indicates the region in the $\{a, b, L\}$ parameter space corresponding to a rate $\Gamma = 10^9$ s^{-1}. All lengths are expressed in nanometers, and the device is made up of Al dots and intrinsic Si acting as an insulator. In the region below the surface, the rate is larger. The continuous lines indicate the operating temperature T_{max}. Reprinted with permission from L. Bonci, G. Iannaccone, and M. Macucci, *J. Appl. Phys.* **89**, 6435 (2001). Copyright 2001, American Institute of Physics.

6.4. Simulation of Clocked and Nonclocked Devices

In order to verify some of the assumptions made in the previous sections, and to assess the actually achievable operating frequency for circuits, we resort to numerical simulation. With a Monte Carlo technique we have studied a chain made of cells such as the one shown in Fig. 6.1 and also more complex circuits, such as a majority voting gate. We were able to compare operation of circuits based on the relaxation to the ground state with that of circuits based on an adiabatic clocking scheme.

The minimum switching time for a clocked array resulted to be much larger than the one obtained from simple time constant considerations. We will discuss later how this can be justified by accurately examining all the processes connected to switching. Before introducing the simulation technique, we need to choose system parameters. From the previously described model we can obtain the values of capacitance and of tunneling resistances for a cell, once the materials and device sizes have been specified. We would also like to compare our theoretical forecast with experimental results, and to extend predictions to circuits more complex than those fabricated so far. We thus choose two different parameters sets: the first relative to actually performed experiments on simple QCA devices,[8] and the second derived from the theoretical analysis of the previous sections. As far as the parameters derived from experiments are concerned, for the nonclocked case we have: $C_o = 400$ aF, $C_v = 288$ aF, $C_c = 88$ aF, $C^* = 1$ aF, $R_T = 200$ kΩ; while for the clocked case we select $C = 420$ aF, $C_1 = 300$ aF, $C_2 = 25$ aF, $C_3 = 80$ aF, $C_4 = 200$ aF, $C_c = 50$ aF, and $R_T = 200$ kΩ. For the theoretical sets of parameters we make choices that represent a trade-off between miniaturization, efficiency, and technical feasibility. We considered cells consisting of Si with embedded cubic 10 nm Al dots, inter-cell distance $a = 2$ nm and intra-cell distance $b = 5$ nm. This leads to the following set of parameters for the nonclocked case: $C_o = 5.3$ aF, $C_v = 5.3$ aF, $C_c = 2$ aF, $C^* = 0.1$ aF, $R_T = 200$ kΩ; while for the clocked case we get $C = 5.3$ aF, $C_1 = 1.2$ aF, $C_2 = 1.2$ aF, $C_3 = 3.57$ aF, $C_4 = 1.48$ aF, $C_c = 1$ aF, and $R_T = 200$ kΩ. From Table 6.3, the maximum operating temperature of such cells is 5.9 K and the transition rate is $\Gamma = 2.1 \times 10^9$ s^{-1}.

Once the tunneling resistance R_T is known, it is possible to write the transition rate between two configurations which differ by an energy Δ and at an arbitrary temperature[7,9]

$$\Gamma = \frac{1}{e^2 R_T} \frac{\Delta}{1 - e^{-\frac{\Delta}{k_B}T}} \quad , \tag{8}$$

where e is the electron charge, k_B the Boltzmann constant and T the temperature.

To obtain an appropriate clock sequence we need to know the maximum and minimum values of V_c, namely the values that correspond to the locked and null states. They depend on system parameters and we have determined them from the energies of the three semicell configurations. We set the null state at the value of V_c for which the $[0, 1, 0]$ configuration energy is $10 \ k_B T$ lower than the minimum energy between the $[1, 0, 0]$ and $[0, 0, 1]$ configurations. Analogously, the minimum of V_c has been chosen as the one for which the energy of the $[0, 1, 0]$ configuration is $10 \ k_B T$ higher than the maximum energy between the $[1, 0, 0]$ and $[0, 0, 1]$ configurations.

6.4.1. *QCA circuit simulator*

For the numerical simulation we used a Monte Carlo code specifically adapted to deal with clocked single-electron devices. Simulation has been performed with a semi-classical approach, following the orthodox Coulomb blockade theory.[9] Quantum effects are indeed negligible at the dimensional scale we are considering, because confinement energies are much smaller than the electrostatic interaction energy.

We have implemented a semi-static Monte Carlo approach, i.e. the voltages from the external sources vary slowly compared to the evolution of the system. We subdivide the simulation time into a number of elementary steps, for each of which we compute the free energy variation Δ for all possible single-electron transitions (considering the current value of the voltages applied to the clock electrodes). From the free energy variations we can obtain the corresponding probability rates using Eq. (8).

After computing all the transition rates, the actual transition is chosen randomly, with a relative probability proportional to its rate, and it is completed with the corresponding variation of dot occupancy. Then the time is increased by the chosen step, new values for the voltage sources are computed, and the procedure is repeated until the end of the simulation interval is reached.

In this description, we did not take into consideration the cotunneling effect, i.e. the simultaneous tunneling of more than one electron. Usually this effect is much less likely than single electron events, but can become important when no single-electron transition are energetically allowed.

The simulation of cotunneling presents many theoretical and numerical difficulties, but it has to be taken into account, because in some operat-

ing conditions it plays an important role, in particular for the evolution
of the nonclocked scheme. We have considered only the first order contribution, i.e., the simultaneous tunneling of two electrons. The two-electron
cotunneling probability can be written as[10]

$$\Gamma = \frac{\hbar}{12\pi e^4} \frac{1}{R_T^{(1)}} \frac{1}{R_T^{(2)}} \left[\frac{1}{\Delta^{(1)} - \Delta/2} + \frac{1}{\Delta^{(2)} - \Delta/2} \right]^2 \tag{9}$$

$$\times \frac{\Delta}{e^{\frac{\Delta}{k_B}T} - 1} [\Delta^2 + (2\pi k_B T)^2)] \, .$$

This is not an expression useful for numerical calculations, because of the
divergences, but an approximate expression can be directly applied to numerical approaches:

$$\Gamma = \frac{16}{3} \frac{\hbar}{\pi} \frac{C^2}{e^8 R_T^2} \frac{\Delta}{e^{\frac{\Delta}{k_B}T} - 1} [\Delta^2 + (2\pi k_B T)^2)] \quad , \tag{10}$$

where R_T and C are the mean values of the tunneling resistances and of
the capacitances, respectively, involved in the cotunneling event. This is a
coarse approximation, which is however sufficient to provide an order of
magnitude estimate of the effect.

6.4.2. *Simulation strategy*

We have focused on the analysis of two different circuits: a simple linear
chain made up of six QCA cells and a majority voting gate made up of
eight QCA cells. Schematic diagrams for these two circuits are presented in
Figs. 6.10–6.12, where we introduce both the nonclocked and the clocked
version of the chain, and the circuit diagram for the nonclocked version of
majority voting gate (since the clocked version is more complex, but is easily
derivable from the schematic diagram of the clocked linear chain). In these
figures, we also sketch the time evolution of the applied potentials. In the
nonclocked case the external gates only enforce switching of the first cell,
while in the clocked case, external gate potentials perform the adiabatic
switching of the cells.   Depending on the presence of the external clock,
we need to change the simulation strategy. Without the clock, we adhere
to the ground-state computation paradigm, and thus we need to wait for
circuit relaxation before accepting the output. This implies, for example
for the chain, that the simulation starts from the ground state represented
by all the cells in the same logical condition, then the first cell switches
and we follow the time evolution, until the new ground state is achieved.

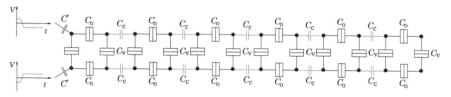

Fig. 6.10.   Chain made up of six nonclocked cells. The first cell to the left is coupled to external leads via the capacitors C^*. The voltage variations which enforce the state of the first cell are sketched next to such capacitors. Reprinted with permission from L. Bonci, M. Gattobigio, G. Iannaccone, and M. Macucci, *J. Appl. Phys.* **92**, 3169 (2002). Copyright 2002, American Institute of Physics.

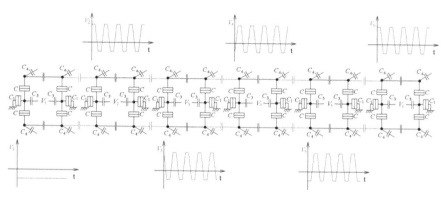

Fig. 6.11.   Chain made up of six clocked cells like the one represented in Fig 6.1. The external voltages which drive the cells are labeled as V_1, V_2 $\cdots$ V_6 and their time behavior is sketched below and above the circuits diagram. The value of the capacitors which connect the cells is $C_c = 1$ aF. Reprinted with permission from L. Bonci, M. Gattobigio, G. Iannaccone, and M. Macucci, *J. Appl. Phys.* **92**, 3169 (2002). Copyright 2002, American Institute of Physics.

To evaluate the reliability of operation and to compute the average time needed to attain the correct logical output, we must repeat the simulation many times, in order to collect appropriate statistics.

A completely different situation is represented by the the clocked case, in which the switching time is enforced from the outside. In order to analyze circuit operation, in this case we follow the time evolution of each single cell, verifying that its logical state is correct in the appropriate time intervals. We do not need to perform ensemble averages, because time statistics are obtained observing the system over a large number of clock cycles.

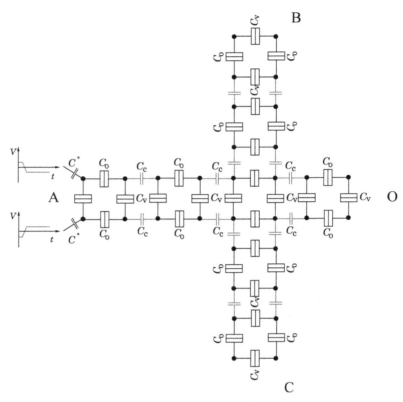

Fig. 6.12. Majority voting gate made up of nonclocked cells. The first cell to the left is coupled to external leads via the capacitors C^*. The voltage variations which cause switching of the first cell are sketched to the left. The other two input cells, at the top and bottom of the vertical arm, are kept in a constant logical state via appropriate external gates (not shown), during the whole time evolution. Reprinted with permission from L. Bonci, M. Gattobigio, G. Iannaccone, and M. Macucci, *J. Appl. Phys.* **92**, 3169 (2002). Copyright 2002, American Institute of Physics.

6.4.3. *Binary wire simulations*

In this section we study the operation of the binary wires of Figs. 6.10 and 6.11. We start with the simulation of devices with experimental parameters. In Fig. 6.13 we show the time evolution of the six cells of the chain of Fig. 6.10. The logical states are indicated with the symbols 1, x and 0, where 1 and 0 correspond to the two valid polarization states, with the two electrons aligned along one of the diagonals, while x indicates all the other cell states that are not associated with a logical value. We initialize

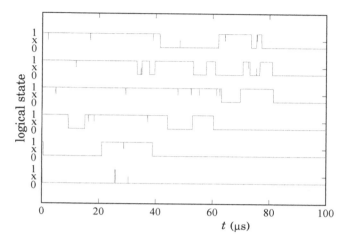

Fig. 6.13. Time evolution of the logical states of the six cells of a QCA chain in the nonclocked case. The values of the parameters are C_o=400 aF, C_v=288 aF, C_c=88 aF, C^*=1 aF, and R_T = 200 kΩ. Reprinted with permission from L. Bonci, M. Gattobigio, G. Iannaccone, and M. Macucci, *J. Appl. Phys.* **92**, 3169 (2002). Copyright 2002, American Institute of Physics.

the simulation from the chain ground state, with all the cells in state 1, and, after a short time, we switch the first cell state to 0 by means of the external voltages. After switching of the first cell, the system evolves toward relaxation into the new ground state, which is the one with all the cells in the logical state 0. It is obvious and apparent from the figure that the relaxation process is not smooth and evolves with random jumps among the various possible chain states. The energy separation among many of these states is indeed very small, and even states of higher energy can be reached, due to thermal fluctuations.

The time needed to complete relaxation to the ground state is a statistical quantity which we have evaluated by averaging over several realizations. Results are shown in Fig. 6.14 as a function of temperature. We note that, due to the increase in tunneling rate and to the contribution of thermal excitations to the escape from metastable states, $< t_{rel} >$ decreases as temperature is raised. This effect is actually limited by the impossibility of approaching a stable ground state above a certain temperature. This disruption of operation, due to thermal fluctuations overcoming the splitting between the first excited state and the ground state, has been indicated in the figure with a hatched region. Indeed, there are already indications of deteriorating operation before actually entering this region. An abrupt rise

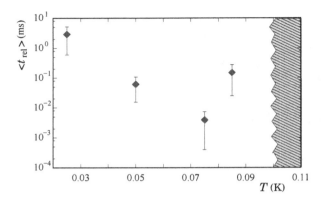

Fig. 6.14.   Average relaxation time $< t_{rel} >$ as a function of temperature. In the dashed region the operation is completely disrupted due to thermal fluctuations. The values of the parameters are C_o=400 aF, C_v=288 aF, C_c=88 aF, C^*=1 aF, and $R_T = 200$ kΩ. Reprinted with permission from L. Bonci, M. Gattobigio, G. Iannaccone, and M. Macucci, *J. Appl. Phys.* **92**, 3169 (2002). Copyright 2002, American Institute of Physics.

of the relaxation time and a broadening of its statistical distribution can be observed, for example, for $T = 0.85$ K.

For the assessment of circuit performance, the average relaxation time is not the only relevant quantity. We must also take into consideration the long tails in the distribution of the relaxation times: we cannot consider the calculation as completed until we have a reasonable certainty that the tail of the distribution has decayed down to a negligible value. This implies that $< t_{rel} >$ is actually too conservative an estimate. This problem is illustrated in Fig. 6.15, where we show the distribution of the relaxation time for T=0.05 K. Although metastable states do not play a major role in a simple circuit such as the one considered, we believe that the long tails are due to the existence of this long surviving excited states.

Another interesting point is the relevance of cotunneling events in the nonclocked architecture. They acquire a growing importance as temperature is decreased: if we define R_{ct} the ratio of the number of cotunneling events to that of tunneling events, we get, for the case we are currently considering, that at T=0.025 K cotunneling is the dominant process, being $R_{ct} = 35$, while $R_{ct} = 1.4 \times 10^{-2}$ at T=0.075 K.

Moving to the other set of parameters, which we have derived from the theoretical model of the previous sections, we find a qualitatively similar behavior, although with lower temperatures and a faster evolution. In Figs. 6.16–6.17 the results for this case are presented: we observe that the

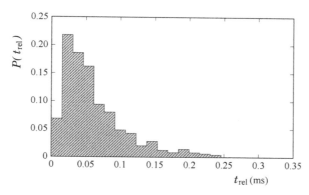

Fig. 6.15. Distribution of the switching time $< t_{rel} >$ for the unclocked chain of Fig. 6.10, for $T=0.05$ K. The values of the parameters are $C_o=400$ aF, $C_v=288$ aF, $C_c=88$ aF, $C^*=1$ aF, and $R_T = 200$ kΩ. Reprinted with permission from L. Bonci, M. Gattobigio, G. Iannaccone, and M. Macucci, *J. Appl. Phys.* **92**, 3169 (2002). Copyright 2002, American Institute of Physics.

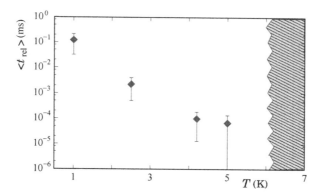

Fig. 6.16. Average relaxation time $< t_{rel} >$ as a function of temperature. In the dashed region the operation is completely disrupted, due to thermal fluctuations. The values of the parameters are $C_o=5.3$ aF, $C_v=5.3$ aF, $C_c=2$ aF, $C^*=0.1$ aF, and $R_T = 200$ kΩ. Reprinted with permission from L. Bonci, M. Gattobigio, G. Iannaccone, and M. Macucci, *J. Appl. Phys.* **92**, 3169 (2002). Copyright 2002, American Institute of Physics.

average relaxation time is now shorter and the operating temperature is higher, but there is no significant change in the qualitative picture. We still find a wide distribution of the relaxation time, with long tails. The improved performance is simply a result of the size reduction, which involves smaller capacitances and higher voltage imbalance, leading in turn to an increased tunneling rate and a decreased sensitivity to thermal fluctuations.

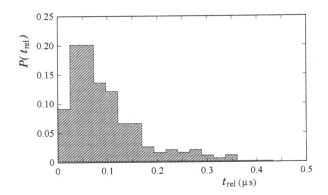

Fig. 6.17. Distribution of the relaxation time $< t_{rel} >$ for the nonclocked chain of Fig. 6.10, for T=4.2 K. The values of the parameters are C_o=5.3 aF, C_v=5.3 aF, C_c=2 aF, C^*=0.1 aF, and R_T = 200 kΩ. Reprinted with permission from L. Bonci, M. Gattobigio, G. Iannaccone, and M. Macucci, *J. Appl. Phys.* **92**, 3169 (2002). Copyright 2002, American Institute of Physics.

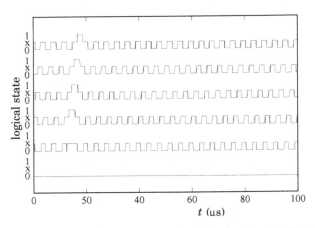

Fig. 6.18. Time evolution of the logical states of the six cells of a QCA chain in the clocked case. The capacitor values are: C=420 aF, C_1=300 aF, C_2=25 aF, C_3=80 aF, C_4=200 aF, C_c=50 aF. The resistance of the tunneling junctions between dots is R = 200 kΩ, while the junctions which connect the dots to the ground have a resistance of $10^{10}\Omega$. Reprinted with permission from L. Bonci, M. Gattobigio, G. Iannaccone, and M. Macucci, *J. Appl. Phys.* **92**, 3169 (2002). Copyright 2002, American Institute of Physics.

In the clocked architecture of Fig. 6.11, the switching rate is enforced by the external clock signal, thus we do not need to look for a relaxation time of the entire circuit, but, rather, to assess whether relaxation to the local instantaneous ground state is fast enough to allow the system to follow the evolution caused by the clock. In Fig. 6.18, we report a sample evolution obtained for the choice of theoretical parameters: within the reported time segment only a single propagation error can be noticed. We do not need to take into consideration cotunneling in this case, because barrier modulation always makes direct transitions possible and thus correlated many-electron tunneling is never relevant.

The reliability of the circuit can be evaluated by investigating many different realizations including a large number (250) of clock cycles. We have monitored the evolution of the second and of the sixth and last cell of the chain. We have estimated the probability that such cells are in the state corresponding to the expected evolution. In particular, we have computed the probability of correct output P_{CO} over all the considered clock cycles. The logical state can be said to be correct if, during the time interval which corresponds to the locked condition, the cell is, on the average, in the expected logical state. We specify "on the average," because, except for extremely low temperatures, we can see the cell switching back and forth between the two states within a single clock phase.

In order to provide a precise definition of *correct logical state* for a cell, we addressed the actual measurement process, which consists in averaging the cell state over the locked time phase. A cell is said to be in the correct state if the correct configuration is found in that cell for more than a certain fraction of the time interval corresponding to the clock phase. We choose such a fraction to be 90%.

Following this approach, we have obtained the results reported in Figs. 6.19–6.20, where the percentage of correct output P_{CO} is reported as a function of the clock period τ. The behavior of the last cell is represented with solid curves, while the dashed curves refer to the second cell. Two values of the temperatures (T=0.01 K (a) and T=0.025 K (b)) have been taken into consideration. Figure 6.19 is a clear example of how circuit operation worsens as clock frequency and temperature are increased. Another apparent result is that the probability of a cell being in the correct state decreases as we move along the chain. This is the consequence of the increase in the overall error probability related to the increase of the number of intermediate cells,[11] as discussed in Chapter 4.

The temperature range for our analysis has been chosen considering that below $T = 0.01$ K thermal effects can be neglected, for our parameter choice, and above $T = 0.025$ K, thermal fluctuations determine a complete failure of the operation of the circuit.

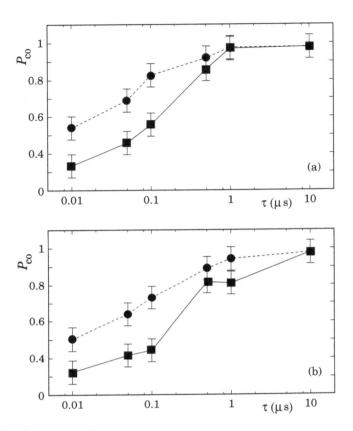

Fig. 6.19.   Probability of correct operation (P_{co}) for the second and the last cell in the clocked chain of Fig. 6.11 as a function of the clock period. The solid curves refer to the last cell, while the dashed ones refer to the second cell. We have considered two different temperatures: T=0.01 K (a) and T=0.025 K (b). Adapted with permission from L. Bonci, M. Gattobigio, G. Iannaccone, and M. Macucci, *J. Appl. Phys.* **92**, 3169 (2002). Copyright 2002, American Institute of Physics.

With the speculative parameter set we have obtained the results shown in Fig. 6.20 for three temperatures: 1 K (a), 2.5 K (b), 4.2 K (c). With this choice, successful operation can be achieved up to 2.5 K and to a clock frequency of 10 MHz.

These results for the maximum achievable clock rate are orders of magnitude smaller than the simple estimate corresponding to the reciprocal of the time constant obtained from the product of the tunneling resistance times the capacitance seen by the nodes. The typical minimum clock pe-

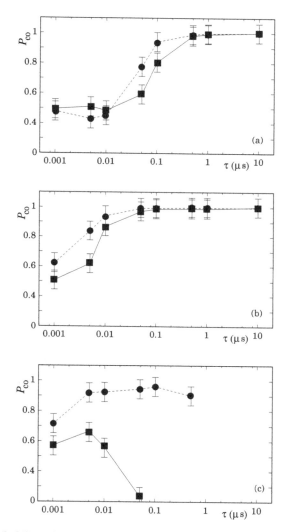

Fig. 6.20. Probability of correct operation (P_{co}) for the second and the last cell in the clocked chain of Fig. 6.11 as a function of the clock period. The solid curves refer to the last cell, while the dashed ones refer to the second cell. We have considered three different temperatures: T=1 K (a), T=2.5 K (b) and T=4.2 K (c). Adapted with permission from L. Bonci, M. Gattobigio, G. Iannaccone, and M. Macucci, *J. Appl. Phys.* **92**, 3169 (2002). Copyright 2002, American Institute of Physics.

riod evaluated from the RC constant is around 10^{-12} s, while our Monte Carlo calculations yield a value of the order of 10^{-8} s. It is possible to

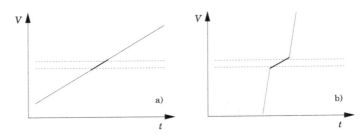

Fig. 6.21. Potential of the central dot vs. time for the clock ramp used in the simulations, (a), and potential of the central dot vs. time for a clock ramp modified to achieve a minimum clock period (b). The dashed lines represent the potential of the two outer dots and the thicker segment indicates the active region. Reprinted with permission from L. Bonci, M. Gattobigio, G. Iannaccone, and M. Macucci, *J. Appl. Phys.* **92**, 3169 (2002). Copyright 2002, American Institute of Physics.

work out an intuitive argument justifying our results: let us consider how an electron on the central dot of a semicell moves into the outer dot with the lower potential. This tunneling process is possible in a time interval somewhat longer (as a result of thermal activation) than the one within which the potential of the central dot is intermediate between those of the top and bottom dots. Such an interval is indicated in bold in Fig. 6.21(a). An estimate of this time interval can be performed starting from the value of the voltage imbalance existing between the two outer dots when an excess electron lies in one of the external dots of the other half of the same cell, i.e. in the presence of a driving action from the other semicell. For our choice of parameters this imbalance is about 3 mV, and, taking into account the linearity of the ramp, we can compute an approximate average electron tunneling rate between the central dot and the outer dot in the condition with the central dot potential lying in the middle between those of the outer dots. In such a situation, across the 200 kΩ tunneling junction we have a voltage of 1.5 mV, leading to a current of 7.5 nA. Such a current is equivalent to an electron flux of 46.82×10^9 electrons/s: the average time an electron takes to tunnel through the junction is the reciprocal of the flux, thus 21.36×10^{-12} s. We must also consider that, in order to achieve proper operation, we choose the intercell coupling to be 5 times smaller than that within a cell, thus the time needed for completion of events triggered by a nearby cell is around 107×10^{-12} s.

Furthermore, we must include the fact that, in order to be reasonably sure that the electron tunneling event actually takes place, we need to allow an active time at least ten times longer than the average tunneling time.

This increases the previous estimate to 1.07×10^{-9} s, and another factor of ten has to be added if we assume linear clock ramps and recall that, to guarantee that the lock condition is effective, the voltage span in a clock cycle must be at least ten times the amplitude of the active interval. To complete the calculation, we must consider that 4 ramps are needed to complete a clock cycle, thus the estimate for the minimum clock period becomes 42.8×10^{-9} s, comparable with the values yielded by our Monte Carlo simulation.

There is room for some improvement, by means of a careful optimization of the parameters. For example, one of the most promising strategies would consist in changing the clock waveforms: with a clock signal such as the one depicted in Fig. 6.21(b), either slowly growing or constant during the active time and rapidly varying otherwise. It would, however, be difficult to supply such a waveform to all cells without significant distortion, and the maximum slope of the clock signal would in any case be limited by the requirement of adiabatic operation.[3]

6.4.4. *Operation of a logic gate*

Let us now move on to the simulation of an actual logic gate, in particular a majority voting gate, whose output, as previously discussed, is in the logical state corresponding to the majority of the inputs.

In Fig. 6.12 a nonclocked implementation of this circuit is presented; its clocked counterpart can be obtained by replacing the standard 4-dot cells with the 6-dot clocked cell of Fig. 6.1.

We start from the investigation of the nonclocked version. The initial condition is assumed to be with all cells in the same state, corresponding to a ground state configuration. Then the state of the input cell A is switched, while keeping the other inputs in the same logical state. The polarization change propagates along the circuit until a new ground state is reached. We compute the relaxation time $< t_{rel} >$ by averaging over a large number of realizations. We notice that, although the simplest way to reach a correct output is to propagate the cell polarization along the horizontal arm, with no change in the vertical arms, this is not the only path followed by the evolution of the system. We have noticed that, especially when temperature is increased, correct propagation along the horizontal arm can occur, while, at the same time, the vertical arms are in an incorrect state. This means that the system, even if it is supplying a correct output, has not reached the ground state. We have assumed these conditions as positive events,

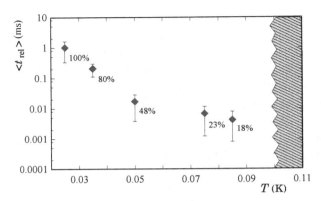

Fig. 6.22.   Average relaxation time $< t_{rel} >$ as a function of temperature for the operation of a nonclocked majority voting gate. Reprinted with permission from L. Bonci, M.
Gattobigio, G. Iannaccone, and M. Macucci, *J. Appl. Phys.* **92**, 3169 (2002). Copyright
2002, American Institute of Physics.

since this is what would be done in a real experiment, in which only the
output state is monitored. For the sake of comparison, we also monitored
the occurrence of the real ground state, indicating its relative frequency in
the figure with the numbers close to the data points (see Fig. 6.22).

Comparing this last result with Fig. 6.14, where the single chain relaxation time is shown, we notice that for the majority voting gate we obtain
a faster relaxation, at least for low temperatures. This was expected, because the distance between the input cell A and the output cell O is shorter
than the length of the chain. However, we have a larger error probability,
which can be attributed to the greater circuit complexity, which implies in
turn a larger number of states very close in energy. This increased complexity is reflected in a reduced reliability, as shown by the fast increase of
the probability of propagation with correct output but non-relaxed vertical
arms.

For the clocked majority voting gate we can adopt the same strategy
used for the clocked chain, namely we monitor the output cell O and, by
averaging its value over the proper portion of the clock period, we are able
to obtain the probability of correct operation that is shown in Fig. 6.23
as a function of clock period for two different temperatures. By comparing Fig. 6.23 and Fig. 6.20, we notice that, for the majority voting gate,
P_{CO} is less dependent on temperature, but shows an overall increased error
probability.

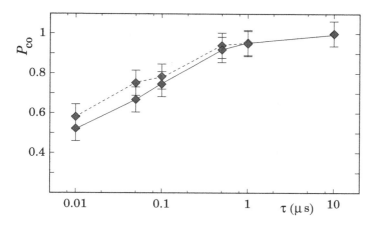

Fig. 6.23. Probability of correct operation (P_{CO}) for the output cell in a clocked majority voting gate, as a function of clock period. We have considered two different temperatures: T=1 K (dashed line) and T=2.5 K (solid line). Reprinted with permission from L. Bonci, M. Gattobigio, G. Iannaccone, and M. Macucci, *J. Appl. Phys.* **92**, 3169 (2002). Copyright 2002, American Institute of Physics.

6.5. Discussion

We have discussed the tunneling rates between dots, via a quasi-analytical formula, for typical material parameters and device geometries, in order to evaluate the region in the parameter space that allows operating circuits with a clock frequency of 10 MHz.

The most promising implementations seem to be the ones based on metal islands embedded in intrinsic semiconductors, because they provide a large electron density and, at the same time, low barrier heights, which leads to a reasonable tunneling rate. In particular, focusing on a six-dot Al-Si cell, we have developed a Monte Carlo simulator to test the validity of our assumptions and to compare the performance of nonclocked and clocked QCA architectures. Our results show that the latter is better suited for computational purposes, because it allows faster and more reliable operation.

The nonclocked architecture is based on relaxation to the ground state, which is a statistical process with a broad time distribution of relaxation times. Moreover, the distribution presents long tails which reveal the existence of metastable intermediate states. The clocked architecture, on the other hand, is characterized by a computation evolving according to an external clock, which allows a much faster and more reliable operation, and enables direct control of the data flow. The limitation on clock rate due to the limited value of the tunneling rate is the main constraint to be considered and we have shown that a reasonable device geometry (although not yet within the reach of present fabrication technologies) allows operation of a QCA chain and of a majority voting gate at a clock period of a few microseconds and at a temperature of a few kelvins. A heuristic argument, considering the details of cell switching, has enabled us to explain why the simple estimation of the switching speed based on RC time constants leads to an error of several orders of magnitude.

Thus the clocked architecture has superior performance with respect to the basic nonclocked QCA circuits, but distribution of the clock introduces very significant complications in circuit design and fabrication, and large switching speeds cannot, unfortunately, be attained.

References

1. C. S. Lent, P. D. Tougaw and W. Porod, *J. Appl. Phys.* **74**, 3558 (1993).
2. R. Landauer, *Ultimate Limits of Fabrication and Measurement*, M. E. Welland, Ed., Dordrecht Kluwer (1994).
3. C. S. Lent and P. D. Tougaw, *Proc. IEEE* **85**, 541 (1997).
4. A. N. Korotkov, *Appl. Phys. Lett.* **67**, 2412 (1995); K. K. Likharev and A. N. Korotkov, *Science* **273**, 763 (1996).
5. G. Tóth and C. S. Lent, *J. Appl. Phys.* **85**, 2977 (1999).
6. N. W. Ashcroft and N. D. Mermin, *Solid State Physics*, Holt, Rinehart and Winston, New York (1976).
7. M. H. Devoret and H. Grabert, in *Single Charge Tunneling*, Plenum Press, New York (1992), p. 12.
8. A. O. Orlov, I. Amlani, G. Toth, C. S. Lent, G. H. Bernstein, G. L. Snider, *Appl. Phys. Lett.* **74**, 2875 (1999); I. Amlani, A. O. Orlov, G. Toth, G. H. Bernstein, C. S. Lent, G. L. Snider, *Science* **284**, 289 (1999); I. Amlani, A. O. Orlov, R. K. Kummamuru, G. H. Bernstein, C. S. Lent, G. L. Snider, *Appl. Phys. Lett.* **77**, 738 (2000); A. O. Orlov, I. Amlani, R. K. Kummamuru, R. Ramasubramanian, G. Toth, C. S. Lent, G. H. Bernstein, G. L. Snider, *Appl. Phys. Lett.* **77**, 295 (2000).
9. D. V. Averin and K. K. Likharev, in *Mesoscopic Phenomena in Solids*, edited by B. L. Altshuler, P. A. Lee, and R. A. Webb, Elsevier, Amsterdam, (1991).
10. L. R. C. Fonseca, A. N. Korotov, K. K. Likharev, A. A. Odinstov, *J. Appl. Phys.* **78**, 3238 (1995).
11. C. Ungarelli, S. Francaviglia, M. Macucci, G. Iannaccone, *J. Appl. Phys.* **87**, 7320 (2000); M. Macucci, G. Iannaccone, S. Francaviglia, B. Pellegrini, *Int. J. Circuit Theory and Applications* **29**, 37 (2001); C. S. Lent, P. D. Tougaw and W. Porod, *Proocedings of the workshop on Physics and Computing*, 17-20 November 1994, Dallas, TX, p. 1.

12. M. Governale, M. Macucci, G. Iannaccone, C. Ungarelli, J. Martorell, *J. Appl. Phys.* **85**, 2962 (1999).
13. R. P. Coburn, M. E. Welland, *Science* **287**, 1466 (2000).

CHAPTER 7

Implementation of QCA Cells with SOI Technology

Freek E. Prins[a], Christof Single, Gregor Wetekam and Dieter P. Kern

Institut für Angewandte Physik
Universität Tübingen
Auf der Morgenstelle 10
D-72076 Tübingen, Germany

Massimo Macucci, Luca Bonci, Giuseppe Iannaccone
and Mario Gattobigio[b]

Dipartimento di Ingegneria dell'Informazione
Università di Pisa, Via Caruso 16, I-56122 Pisa, Italy

7.1. Advantages of the SOI Material System

Several groups have reported single electron charging effects in silicon on insulator (SOI) material, with differences mainly in the approaches used to realize the tunneling barriers. The first experiments were reported by Ali *et al.*[16] in which the tunneling barriers were not obtained by means of lateral constrictions but by thinning a 40 nm thick Si layer locally, down to 10 nm. A Coulomb blockade region of 1 mV was observed at a temperature of 0.3 K. Thermal oxidation to reduce the size of lateral constrictions connecting a small island with leads was introduced by Leobandung *et al.*[12] In these experiments the islands were undoped and charging effects up to 110 K were demonstrated. A stress dependent thermal oxidation was used by Fujiwara *et al.*[17] who realized a multiple island structure and demonstrated a single electron memory effect on satellite Si islands. Another approach

---

[a]Current address: ZMD, Grenzstrasse 28, D-01109 Dresden, Germany
[b]Current address: Scuola Normale Superiore, Piazza dei Cavalieri, I-56126 Pisa, Italy

for the realization of tunneling barriers relies on the statistical distribution of dopants.[18] The potential along a long wire will fluctuate and potential valleys can act as islands. In this chapter we will describe experiments on structures containing lateral constrictions as well as a high dopant concentration. The electrical characterization will lead to a better understanding of single electron charging effects observed in SOI nanostructures.

If devices are to be operated at significantly higher temperatures than 4 K, it is clear that extremely small capacitors and sufficiently transparent tunneling barriers are to be combined. Therefore all parameters influencing the tunneling process as well as the capacitance should be considered. Let us first consider the tunneling probability T of a particle with mass m through an energy barrier with a height ΔE and a width d:

$$T \sim e^{-2\frac{d}{\hbar}\sqrt{2m\Delta E}} . \tag{1}$$

The tunneling probability depends exponentially on the width of the barrier and only for small values of d significant tunneling can be expected.

At the same time, the capacitance should be very small. As an example, a parallel plate capacitor, which is an acceptable model for previously realized devices[1] has a capacitance that can be expressed, as long as the separation L between the plates is much smaller that the transverse dimensions of the plates, with the simple formula:

$$C = \varepsilon\frac{A}{L} , \tag{2}$$

where ε is the permittivity of the dielectric and A is the area of the plates. A small value of the capacitance could in principle be obtained using a large separation L. However, this quantity coincides with the width d of the tunneling barrier, which should be small, in order to obtain a sufficiently large current. These conflicting requirements must be satisfied at the same time for the realization of useful devices based on the Coulomb blockade principle, and they can be achieved simultaneously only with structures in the 10 nm range (so that A can be very small, too). Therefore a dedicated effort for fabrication on a nanometer scale has to be made.

There are other material parameters which strongly influence the tunneling probability, as well as the capacitance. A sufficiently large tunneling probability can be achieved using a small effective mass, a low barrier height and a small width of the barrier. The effective mass for tunneling can undergo significant variations, depending on the material.[2] The barrier height between metals and vacuum is ≈ 5 eV (5.2 eV for Pd [3]) and can be reduced using other media. However, the barrier height should clearly be larger than the thermal energy at the intended operating temperature. The capacitance can be further reduced using a material with a small dielectric constant. As this constant is around 10 for semiconductors, the capacitance can be changed by an order of magnitude using different materials. However, the dielectric constant and the height of the energy barrier cannot be chosen independently.[4] Materials and systems with a small dielectric constant tend to have a large energy barrier and vice versa (see Table 7.1). This means that an improvement consisting in the reduction of the dielectric constant, leading to a smaller capacitance, can be washed out by a reduction of the tunneling probability. In order to compensate for this, the distance between the electrodes should be reduced, increasing the capacitance again. Such dependencies make the choice of material a complex problem.

Table 7.1. Permittivity and barrier height for various media.

	ϵ	E (eV)
Pd/Vacuum/Pd	1	5,2
Al/AlO$_x$/Al	8	1
Ti/TiO$_x$/Ti	24	0,28

One approach for the realization of Coulomb blockade devices is the use of metallic clusters or colloidal particles.[5] The starting point here consists in prefabricated electrodes on the substrate surface. Particles or clusters can be deposited statistically, e.g. from a liquid over the entire sample surface. After drying, some particles will lie between the electrodes and can act as small conducting islands. Electrons can tunnel from the electrodes, through the vacuum onto the islands. As the dielectric constant of vacuum is small, very small capacitances of these islands can be expected. On the other hand, the high electron mass and the high tunneling barrier require a small distance between the electrodes and the particles, of the order of 1 nm. As the positioning of particles during deposition is rather uncontrolled, a random pattern of many particles between the electrodes is obtained. In such structures a clear Coulomb blockade was observed at elevated temperatures, but the results are not well understood and, due to the random pattern of the particles, the results differ significantly from sample to sample. Many other experiments on Coulomb blockade have been made using Al/AlO$_x$/Al junctions.[1,6] First, a thin aluminum layer is evaporated. After the evaporation, a thin AlO$_x$ layer is grown on the Al and subsequently a second Al layer is evaporated. If the overlap between the two Al layers is kept minimal, a very small Al/AlO$_x$/Al capacitor is obtained. In this material system, the height of the tunneling barrier is reduced to 1 eV[7] and the dielectric constant of the oxide increases to 8.[8] With the high effective mass of the electrons in this material system, the width of the tunneling barrier (i.e. the oxide thickness), must be of the order of 1 nm. This increases the capacitance and thus, in order achieve a capacitance of 10 aF, an area of 10 nm^2 is required (using Eq. (2)), which is not feasible with current e-beam lithography. Therefore, these devices are currently only suitable for low-temperature experiments.

Another experiment was performed with a 3 nm thin Ti layer.[9] This layer was locally oxidized under the tip of a scanning tunneling microscope (STM). The height of the barrier in the resulting tunnel junction is 285 meV and the relative permittivity of the oxide is 24. The reduced barrier height allows a somewhat wider barrier. The high permittivity, however, leads to an increase of the capacitance, but this is compensated by the extremely small thickness of the Ti layer, which limits the surface of the electrodes. In this system, Coulomb blockade has been demonstrated at room temperature, but fabrication of a homogeneous layer of such a thickness is difficult, because of the formation of grains, and the lithographic technique used is very slow.

Coulomb blockade has also been observed in semiconductor nanostructures. Most well-known are the experiments on III-V semiconductor heterostructures like GaAs/AlGaAs.[10,11] Metal electrodes are defined on the surface, and, by biasing them with a voltage, the 2-dimensional electron gas below the surface can be selectively depleted. In this way, structures and tunneling barriers can be formed, however, since the conducting layer is characteristically about 50 nm below the surface, the lateral resolution of the potential created by the surfaces electrodes is of the same order of magnitude. Therefore, the structures defined in this material system are necessarily relatively large. On the other hand, the small effective mass of the electrons in these semiconductors and the small height of the barrier allow a wide tunneling barrier, so that Coulomb blockade can be observed. Especially the fact that the height and the width of the barrier, as well as the size of the island, can be controlled by means of the voltages on the electrodes, makes this material very suitable for basic studies of fundamental physical effects at low temperatures.

Another approach consists in the fabrication of nanostructures in thin films by means of lithography and etching, as shown in Fig. 7.1.[12] An island is fabricated between two electron reservoirs, source and drain, and is con-

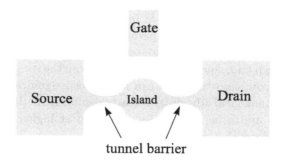

Fig. 7.1. Schematic structure of a Coulomb blockade device in a thin film. The island is connected to the source and the drain by small constrictions, which form tunneling barriers.

nected to them with small constrictions. If the width of these constrictions is smaller than the Fermi wavelength λ_f of the electrons in the reservoirs, the constrictions will form a tunneling barrier. In metals, the Fermi wavelength is very small: in the case of gold, $\lambda_f = 0.5$ nm. The fabrication of constrictions on this scale is beyond the possibilities of any lithographic technique. However, the Fermi wavelength strongly depends on carrier density and is therefore much larger in semiconductors (e.g. 20 nm in Si at a carrier density of 10^{18} cm^{-3}). The achievement of such dimensions in semiconductors is feasible. Since in this case electrons still tunnel through semiconductor material, a small electron mass compared to vacuum can be assumed. This allows tunneling through a relatively thick barrier. Furthermore, as most of the material between island and reservoir is removed by etching, the permittivity in this region is effectively lower than that in bulk semiconductor structures. This, together with the thick tunneling barrier, allows the achievement of small capacitances. Furthermore, the height of the tunneling barriers depends on the width of the constrictions and can also be adjusted by changing the carrier density in the structure.

The important result is that the height of the tunneling barrier and the value of the dielectric constant are almost completely decoupled, which makes this approach quite interesting. The transport properties of small III-V semiconductor heterostructures realized by lithography and etching have been studied extensively:[13,14] etch damage and surface effects dominate electron transport in these structures and, as dimensions approach the region of interest, conduction approaches zero.

In recent years however, great progress has been made in the fabrication of thin silicon layers on silicon oxide.[15] In contrast to III-V semiconductors, silicon can be annealed at high temperatures (1000 °C) after etching, and in this way etch damage can be removed. In addition, during annealing an oxide can be grown and, since in this process Si from the sample is used, the dimension of silicon nanostructures can be reduced significantly. All these physical and technological advantages clearly demonstrate that, if Coulomb blockade devices are to be realized for operation at higher temperatures, silicon is a very good choice.

7.2. Fabrication of Si-Nanostructures

The fabrication process for silicon Coulomb blockade devices is schematically shown in Fig. 7.2. The starting material is a thin (20-50 nm thickness) silicon-on-insulator (SOI) layer, in which the insulator is a buried silicon oxide layer. The structures are defined with lithography and created by means of dry etching. Electron beam lithography is used to define the initial silicon structures, as this technique offers the required resolution, and a high degree of flexibility. After Cr-evaporation and lift-off, an etch mask is realized on top of the silicon layer. This mask is transferred into the SOI in a fluorine based anisotropic reactive ion etching process, which was optimized to obtain steep side walls. All structures were over etched into the insulator, as this reduces possible background charges in the vicinity of the structures. Although the electron beam lithography used offers a 10 nm resolution, it is not possible to reliably define the small constrictions which form the tunneling barriers. Therefore we resort to thermal oxidation of the etched structures: during oxidation, silicon is taken from the nanostructures to form a SiO_2 layer, which implies that the width of the nanostructures is reduced in a controllable way. Another advantage of this method is that possible crystal damage induced by the etching process is annealed. Figure 7.3 shows a typical example of a structure.

In order to determine the dimensions of the silicon structures, the oxide was chemically removed in buffered fluoridric acid (HF). The dot diameter is estimated to be 25 nm and the constrictions are 10 nm wide.

7.3. Experiments with the SOI Material System

Coulomb blockade structures in SOI-material were realized, such as those shown in Fig. 7.3. If we assume that the dot can be modeled as half of a spherical capacitor, we can estimate the capacitance of this 25 nm dot to

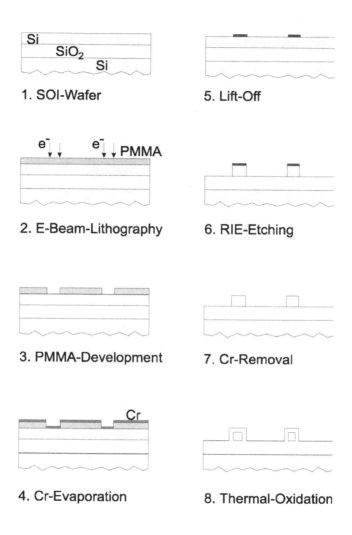

Fig. 7.2.   Process scheme for the realization of Si-nanostructures.

be 30 aF, which roughly corresponds to a Coulomb gap of 5 meV. After fabrication of the contacts, the devices are ready for electrical characterization, in which the substrate is used as a back-gate to influence the carrier

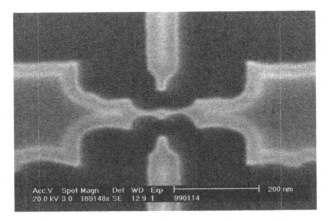

Fig. 7.3. Scanning electron micrograph of a Si-nanostructure. The oxide was removed
with wet chemical etching. Reprinted with permission from R. Augke, W. Eberhardt,
C. Single, F. E. Prins, D. A. Wharam, D. P. Kern, *Appl. Phys. Lett.* **76**, 2065 (2000).
Copyright 2000, American Institute of Physics.

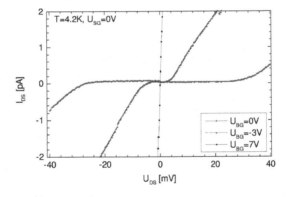

Fig. 7.4. Current-voltage characteristics of a Coulomb blockade device for different
back-gate voltages U_{BG}. With increasing U_{BG} a decrease and subsequent disappearance
of the Coulomb gap can be observed.

density in the entire structure and the side gates are used to adjust the
potential in the dot. Figure 7.4 shows the $I - V$ characteristic of the device
for different back-gate voltages, with grounded side gates. A clear Coulomb
blockade region can be observed, which strongly depends on the back-gate
voltage. At negative gate voltages, the Coulomb gap is much larger than

expected from the geometry of the structure, while at gate voltages higher than 7 V the blockade is completely lifted.

The large Coulomb gap, considering device dimensions, can be explained with the presence of multiple tunnel junctions in series.[18,19] We expect the tunneling barriers and the islands to be not only defined by geometry, but mainly by random fluctuations of the potential as a result of the random location of the dopant atoms. This effect is schematically represented in Fig. 7.5. Here a structure with 4 islands is shown, through which, even if a large voltage is applied, no electron can be transferred, because the voltage drop over each single barrier is insufficient to lift the Coulomb blockade. This leads to the observation of an artificially large Coulomb gap. When the back-gate voltage is changed, the Fermi level, carrier density and Fermi wavelength are changed. Thereby, the size and the number of the islands as well as the width and the height of the tunneling barriers is modified, which leads to the strong dependence of the Coulomb gap on the back-gate voltage. The complete lifting of the Coulomb blockade at large values of U_{BG} can then be explained by the Fermi level raising above the tunneling barriers.

The current through the device depends on the source-drain voltage U_{DS}, as well as on the side gate voltage U_{SG}. For a complete characterization of the structure, the dependence of the conductance as a function of both voltages must be measured systematically. The essential characteristics of the device are shown in Figs. 7.6 and 7.7. For negative side gate voltages U_{SG} the Coulomb blockade cannot be completely lifted. Only for

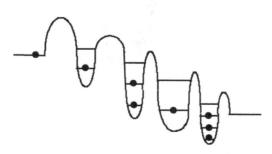

Fig. 7.5.   Schematic drawing of multiple tunnel junctions. No current can flow, although the energy due to the applied voltage is larger than the charging energy of each dot.

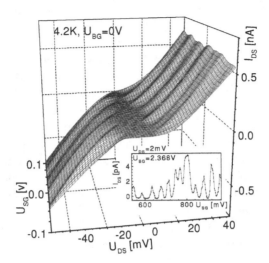

Fig. 7.6.   Current I_{DS} as a function of the side gate voltage U_{SG} and source-drain voltage U_{DS}. At gate voltages above -10 mV periodic lifting of the blockade is possible. In the inset, clear Coulomb oscillations are shown.

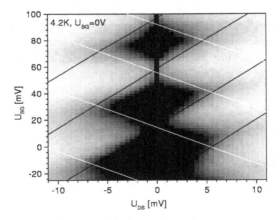

Fig. 7.7.   Logarithmic gray scale plot of I_{DS} versus the side gate voltage U_{SG} and the source-drain voltage U_{DS}. Black areas represent the blockade regime, while bright areas represent the transport regime.

large U_{DS} (> 10mV), stochastic oscillations in the current as a function of U_{SG} are observed, which is an indication of the presence of multiple islands with different sizes and different capacitive coupling to the gates. For $U_{SG} > -10$ mV the blockade is lifted periodically, and rhombically shaped blockade regions are observed. This latter pattern is expected for single dot characteristics. In the inset of Fig. 7.6, periodic oscillations in current are shown, suggesting the presence of a single island. At each current peak, a value of the chemical potential in the dot is located between the Fermi energies of the reservoirs.

There are three possibilities to explain the observation of the single dot characteristics shown in Figs. 7.6 and 7.7. First, at these gate voltages only one dot exists in the structure, which is supported by the fact, that the observed Coulomb gap matches the gap which is expected for this geometry. A second possibility is that there are multiple dots with exactly the same gate capacitance.[20] This, however, is very unlikely in dots formed by random impurity fluctuations. The third possibility is the presence of dots with strongly different gate capacitances.[21] A second dot with a much larger gate capacitance could give rise to a very small peak separation, smaller than thermal broadening. This dot would then always be in a conducting state. However, a dot with such a large gate capacitance is very unlikely, because it should have a size larger than the dimensions of the structure.

From the slopes in Fig. 7.7 and the oscillation period of 42 mV in Fig. 7.6, the constant interaction model[22] yields an estimate for the total capacitance of the island of 32 aF. This is in good agreement with the capacitance calculated earlier from geometry, with the size of the observed Coulomb gap, as well as with the measured temperature dependence. We therefore conclude that for a Coulomb gap of approximately 5 meV the structure consists of a single dot, whose size is close to the geometrically defined region.

In this chapter we shall focus on nanostructures based on the Si/SiO_2 material system. The value estimated for the total capacitance of the structures does not allow the operation of the devices at temperatures significantly higher than 4 K. However, this result must be compared with what is achievable in other material systems. Characteristic values for the dot capacitance are 200-400 aF in GaAs/AlGaAs heterostructures as well as in $Al/AlO_x/Al$ junctions.[1,11] A significant decrease of the capacitance is clearly accomplished with the material system we have chosen.

7.4. Electrical Characterization of Double Dots

As a further step on the way toward implementation of a 4 dot QCA cell, a double dot structure has been fabricated with the described SOI technology, as shown in Fig. 7.8. The structure was completely $5 \cdot 10^{18}$ cm^{-3} n-doped (phosphorous) and the distance between the centers of the dots is 100 nm. The nomenclature of the electrical parameters used in the following experiments is indicated in Fig. 7.8.

First the $I - V$ characteristics were measured, as a function of the back-gate voltage, while both side gates were grounded: clear Coulomb blockade regions were observed. For $V_{\mathrm{bg}} = 0$ V, the blockade voltage is large, which can again be explained by the hypothesis of multiple tunnel junctions. The size of the Coulomb gap decreases with increasing V_{bg}, which can be explained by an increase of the Fermi level in the conducting Si-layer.

In a systematic investigation, the dependence of the Coulomb gap and of the Coulomb oscillations on V_{bg} was determined. For different V_{bg}, $I - V$ curves were measured as a function of the side gate voltage for both side gates V_{sg}, as shown in Fig. 7.9. Both side gates were kept at the same voltage. At a back-gate voltage of 20 V the maximum width of the Coulomb blockade region is about 40 mV. The blockade cannot be lifted by the side gates and no regular structure can be recognized. This is characteristic for multiple tunnel junctions, as was described earlier, due to random impurity fluctuations. As V_{bg} is increased to 36.7 V, the maximum width of the

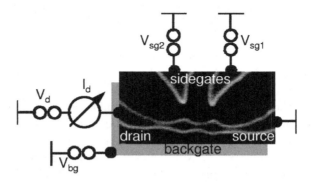

Fig. 7.8.   Experimental set-up for the characterization of a double dot structure. The Si-substrate is used as a back gate electrode. Reprinted with permission from C. Single, R. Augke, F. E. Prins, D. A. Wharam, D. P. Kern, *Appl. Phys. Lett.* **14**, 1165 (1999). Copyright 1999, Institute of Physics.

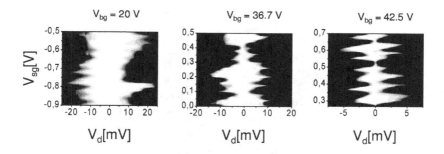

Fig. 7.9. Gray scale plots of the absolute value of the drain current $|I_d|$ vs. drain voltage V_d and side-gate voltage V_{sg} (both side gates are kept at the same voltage) for different back-gate voltages. White corresponds to $|I_d| = 10^{-15}$ A, black to $|I_d| = 10^{-12}$ A. Adapted with permission from C. Single, R. Augke, F. E. Prins, D. A. Wharam, D. P. Kern, *Semicond. Sci. Technol.* **14**, 1165 (1999), Copyright 1999, Institute of Physics Publishing Ltd.

blockade region decreases to 20 mV. The diagram is still irregular but the Coulomb blockade is lifted stochastically. Finally, at a back gate voltage of 42.5 V, the maximum width of the Coulomb gap is reduced to 10 mV, the blockade is lifted periodically, and regular regions appear in the diagram. This is the typical behavior for a structure containing a single dot.

With increasing V_{bg}, the Fermi level is shifted raises above the tunnel barriers and dots will merge. We can thus adjust the number of active dots and the width of the Coulomb blockade region by applying an appropriate back-gate voltage. In the spherical capacitor approximation, we estimate a capacitance per geometrical dot of 10 aF. This corresponds to a maximum width of the Coulomb blockade region of the order of 20 mV, as is the case for $V_{bg} = 36.7$ V, as shown in Fig. 7.9. Therefore, in all successive experiments the structures were biased in this regime and V_{bg} was held fixed.

In the next experiment, the individual influence of the side gates was investigated. When changing the voltages applied to the side gates, the current through the double dots can be easily switched "on" and "off." In Fig. 7.10 the dependence of the source-drain current on the side gate voltages is shown in a gray scale plot. Here bright regions represent higher currents. A clear systematic, regular structure can be observed. These so called honey-comb structures are characteristic for double dot devices.[20,23] This is proof that, although our structures are doped, we can achieve a

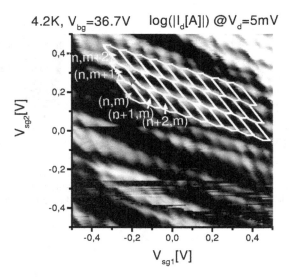

Fig. 7.10.   Logarithmic gray scale plot of the current through the double dot device as a function of the side gate voltages. Bright regions represent higher currents.

situation in which only 2 dots are active. Each hexagonal cell in Fig. 7.10 represents a state where a particular electronic configuration (n, m) is the ground state, with n and m corresponding to the number of excess electrons on the dots. In the interior of the honey-comb cells there is no charge transport through the double dot, due to Coulomb blockade. By changing the voltages of the side gates, the charge configuration of the double dot structure can be changed, crossing the borders between cells.

Each time a region of higher current is passed, the charge in the system is changed by 1 electron. By adjusting the side gate voltages in such a way that a path through one current peak is taken (e.g. from the (n, m) state to $(n+1, m)$, and further to $(n, m+1)$ to end in the (n, m) state), exactly one electron is transferred from the source, to the first dot, to the second dot and finally to the drain. Further, when going from one stable state to another, as indicated by the dark arrow in the charging diagram in Fig. 7.10, the charge on the double dot can be changed from the $(n+1, m)$ to $(n, m+1)$ configuration. During this transition, shown in Fig. 7.11, one electron is transferred from one dot onto the other. The demonstration of this interdot electron transition is an important step toward QCA operation.

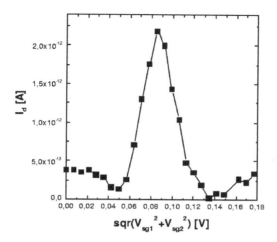

Fig. 7.11. Drain current for a side gate voltage sweep corresponding to an interdot electron transition according to the dark arrow indicated in Fig. 7.10.

From these charging diagrams, capacitances between the side gates and the dots as well as the cross capacitances, can be determined, with the help of a model.[21,24] Here, the electrostatic energy E of the entire structure is calculated as a function of the number of electrons on each dot and of the side gate voltages. The charge configuration of the dots can only change and current can only flow, if the energies for two configurations are equal.

In the schematic diagram of Fig. 7.12 the naming of the capacitances is defined. From the model, we determined gate-dot capacitances $C_{g1d1} = 2.19$ aF and $C_{g2d2} = 1.94$ aF. The cross capacitances C_{g1d2} and C_{g2d1} were significantly smaller, with 1.03 and 0.61 aF, respectively.

7.5. Electrical Characterization of a 4 Dot QCA Cell

For a complete demonstration of cell operation, the first requirement is that a pair of adjacent double dot structures is realized and operated in the double dot regime simultaneously.

A SEM-micrograph of the structure is shown in Fig. 7.13. Assuming that the dots can be described by isolated spherical capacitors embedded in SiO_2, a dot capacitance of 6 aF can be estimated. This would lead to a Coulomb gap of 27 mV. However, external electrodes, especially source

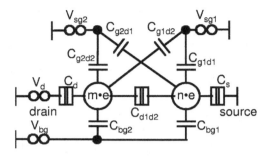

Fig. 7.12. Assumed equivalent circuit. Reprinted with permission from C. Single, R. Augke, F. E. Prins, D. A. Wharam, D. P. Kern, *Semicond. Sci. Technol.* **14**, 1165 (1999), Copyright 1999, Institute of Physics Publishing Ltd.

and drain, are in close vicinity of the dots, and thus a larger capacitance can be expected. The experiments presented above for dots with similar dimensions showed that for each dot a Coulomb gap of 10 meV can be expected.[27,28]

The 4 Si-dots, separated by a distance between centers of 100 nm, form the QCA cell. The device is formed by two double-dot structures, one at the top and one at the bottom of the cell. Each double dot consists of two series connected dots, interconnected and coupled to two electron reservoirs (source and drain) by geometrical tunnel barriers. In order to adjust the potentials of the dots, a side gate is fabricated close to each dot. Furthermore, the Si substrate can be used as a back-gate, influencing the carrier density and the Fermi level in the entire structure. For convenience, in Fig. 7.13 the nomenclature of the gates and dots is reported.

High dopant concentration levels can lead to randomly distributed potential fluctuations.[18,29] These fluctuations can form additional tunnel barriers and the electrically active structure of each dot splits into multiple dots, leading to a multiple tunnel junction regime. Using the back gate, the Fermi energy in the structure can be adjusted. When the Fermi level rises over a barrier, the adjacent dots merge. While continuing this process, eventually two dots are left and the electrically active structure comes close to the desired geometrical structure of the device.[27]

The $I - V$ characteristics of both double dots as a function of the back-gate voltage V_{bg} are shown as a gray scale plot in Fig. 7.14. Dark regions

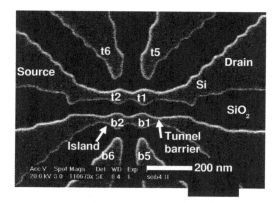

Fig. 7.13. Micrograph of the 4-dot cell in Si, connected by geometrical tunnel barriers (TB) to the reservoirs, source and drain. The nomenclature of the applied voltages for the top and bottom double dot is also shown. Reprinted with permission from C. Single, F. E. Prins, D. P. Kern, *Appl. Phys. Lett.* **78**, 1421 (2001). Copyright 2001, American Institute of Physics.

represent low currents, while brighter regions represent high currents. For both structures, a strong suppression of the current around 0 V is observed, due to Coulomb blockade. At $V_{\mathrm{bg}} = -10$ V we observe a Coulomb blockade region in both structures, of up to 40 mV, which is larger than what was expected for this geometry. Furthermore the blockade could not be lifted completely by the gate. This is, again, due to the formation of multiple tunnel junctions.

At less negative V_{bg}, around -7.5 V, the Coulomb blockade region is reduced to 20 mV, a value consistent with geometry. The blockade can be lifted completely, but not periodically. This is characteristic for a regime where 2 independent dots are active. The blockade is only lifted completely when both dots are in a conducting state simultaneously. This only happens stochastically, because of the different energy level spacing in the dots and the different coupling of the dots to the back-gate.

When the gate voltage is further increased, the width of the Coulomb blockade region is further reduced and a more regular pattern is observed. This can be understood assuming that the Fermi level rises above one of the remaining barriers and only one electrically active island is left. From a standard Coulomb blockade characterization, the capacitance of this remaining single dot is estimated to be 28 aF. This is an upper limit, since when the Fermi level is reduced the electrically active areas will become

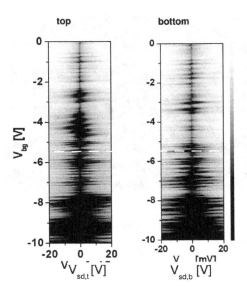

Fig. 7.14.   Simultaneous measurement of I_{SD} for the top (left) and bottom (right) double dot as a function of the back gate voltage V_{bg}. Black indicates $|I_{SD}| < 10^{-14}$ A and white corresponds to $6 \cdot 10^{-12}$ A on a logarithmic scale. The dashed lines indicate the region where double dot behavior can be expected. Adapted with permission from C. Single, F. E. Prins, D. P. Kern, *Appl. Phys. Lett.* **78**, 1421 (2001). Copyright 2001, American Institute of Physics.

smaller, tunnel barriers will become wider, and therefore the actual capacitances in the double dot regime will be reduced. Assuming symmetric tunnel barriers, a tunnel resistance of 80 MΩ can be estimated.

Within the region $-7.5 < V_{\text{bg}} < -5$ V both double dots show a similar behavior, characteristic for a 2 dot system. It is in this regime, specifically at $V_{\text{bg}} = -6$ V, where further systematical measurements have been performed in order to study the properties of the complete four dot structure.

A confirmation of the presence of a 2 dot system in both double dots can be obtained from the charging diagrams. First, the source-drain current $I_{\text{SD,t}}$ in the top double dot is measured as a function of the side gate voltages V_{b5} and V_{b6} at a constant V_{bg} of -6 V. Subsequently, the same experiment at the same V_{bg} is performed for the bottom double dot. Results are shown in the gray scale plots of Fig. 7.15. Here bright regions represent higher currents. For the top dots, see Fig. 7.15(a), a clear systematic regular

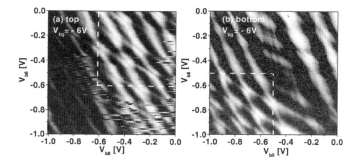

Fig. 7.15. Simultaneous measured charging diagrams for the top (a) and the bottom (b) double dot for V_{bg}=-6 V and $V_{SD,t} = V_{SD,b}$=3 mV. Both diagrams show characteristic honey-comb structures within the regions indicated by the broken lines. Black indicates that $|I_{SD}| < 2 \cdot 10^{-14}$ A and white corresponds to $2 \cdot 10^{-11}$ A on a logarithmic scale. Adapted with permission from C. Single, F. E. Prins, D. P. Kern, *Appl. Phys. Lett.* **78**, 1421 (2001). Copyright 2001, American Institute of Physics.

pattern can be observed, especially for side gate voltages between -0.6 V and 0 V. For the bottom double dot, Fig. 7.15(b), the same regular pattern appears very clearly for gate voltages between -1 V and -0.5 V. These so called honey-comb structures are characteristic of double dot devices.[22,24] This is evidence that, although our structures are doped, we can achieve a situation in which only 2 dots are observed in both double dots at the same V_{bg}. Differences between the electronic configurations in both structures are compensated by means of the offset in side gate voltages. A further indication of the different potential background experienced by each dot is the noise in Fig. 7.15(a). It may be attributed to a change of the electron population of a trap, yielding a change of the potential on the dots. This is equivalent to a shift on the gate voltage axis, as observed in Fig. 7.15(a). The trap should be in close vicinity of the top double dot, since such a noise is not observed for the bottom double dot.

Figure. 7.15 can be interpreted as the charging diagram of the structures. The higher current regions (bright in Fig. 7.15) form the boundaries delimiting regions with a constant electronic configuration. Each time a region of higher current is traversed, the charge in the system or its distribution between the dots is changed by the amount of 1 electron. When using all 4 side gates, we are able to manipulate the number of electrons on each individual dot simultaneously. This is an important achievement towards future operation as a QCA cell.

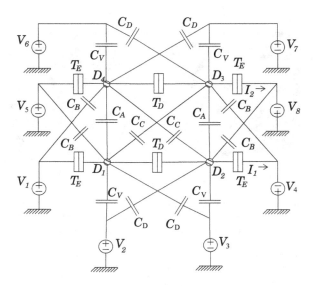

Fig. 7.16. Equivalent circuit diagram for a QCA cell in SOI technology.

An equivalent circuit diagram of the structure, defining the naming for the various capacitances, is shown in Fig. 7.16. For each capacitor the first subscript indicates the function (according to lettering in Fig. 7.16), the second indicates the vertical position (top or bottom) and the third specifies the lateral position (left or right). Capacitors corresponding to the tunnel junctions are indicated with a first subscript "T" and a third subscript "l" (left), "m" (middle) or "r" (right). From the charging diagrams, the capacitances between the gates and the dots and between dots can be estimated. Further, the ratio of the total capacitance of a dot (indicated with two subscripts standing for the vertical and horizontal position) and the interdot capacitance can also be determined.[22,24] The results are shown in Table 7.2. The average capacitance between the dots and their nearest neighboring gate is 1.10 ± 0.04 aF, which is very uniform. This uniformity is a further strong indication that the electrically active structure is close to the geometric Si-structure. Randomly formed dots would result in strongly differing capacitances. The capacitances between the dots and the second-nearest neighboring gates (e.g. between the top right dot and the top left gate) were estimated, too. Since the distances between the gates and the dots, as well as those between the dots of the different double dots are roughly equal, similar capacitances between the dots of the top and the

Table 7.2. Capacitances as estimated from
the the charging diagrams of Fig. 7.15.

top double dot	bottom double dot
$C_{Vtr} = 1.11$ aF	$C_{Vbr} = 1.06$ aF
$C_{Dtl} = 0.53$ aF	$C_{Dbr} = 0.43$ aF
$C_{Vtl} = 1.14$ aF	$C_{Vbl} = 1.07$ aF
$C_{Dtl} = 0.53$ aF	$C_{Dbl} = 0.57$ aF
$C_{tl}/C_{Ttm} = 2.67$	$C_{bl}/C_{Tbm} = 3.11$
$C_{tr}/C_{Ttm} = 2.70$	$C_{br}/C_{Tbm} = 2.85$
$C_{tl}/C_{tr} = 0.99$	$C_{bl}/C_{br} = 1.09$

bottom double dot, e.g. C_{Ar}, can be expected.

7.6. Concept of an Experiment for the Detection of QCA Operation

Based on the successful fabrication of SOI structures containing quantum dots joined by tunnel barriers, we have devised an experimental procedure[25] that should be able to prove QCA operation in an SOI cell without the need of external charge detectors, which would add significant complexity to the structure. Let us consider, for the description that will follow, the equivalent circuit of Fig. 7.16. There are 4 tuning gates, which are biased by the voltage sources V_2, V_3, V_6 and V_7, and are capacitively coupled to the nodes D_1, D_2, D_3 and D_4. The dots belonging to the same pair are connected through the tunneling junctions T_D and couple to the external leads through the junctions T_E. Along the vertical direction, dots are coupled by the capacitances C_A, while the capacitors C_C create diagonal connections in the cell (which may be responsible for severe degradation of QCA operation). To prevent excessive crowding of the diagram, we have not represented the capacitances between each tuning gate and the dots belonging to the opposite pair, which have however been properly included into the simulations.

Let us start from the analysis of the operation of a single pair of dots, D_1 and D_2, assuming that all voltage sources connected to the upper half of the cell are short circuited. The number of electrons on each dot can be varied by tuning the bias voltage applied to the adjustment gates, and, if a voltage imbalance is created between the left and the right lead, a nonnegligible current will flow only if the chemical potentials of the dots lie between those of the leads, i.e. if the Coulomb blockade is lifted. For the experiment we consider only the condition with a very small voltage applied between the input (left) and output (right) leads, therefore this condition will be equivalent to that of having the chemical potentials in the dots lined up with those in the leads.

If we apply voltage ramps of opposite slope (positive to the left gate and negative for the right gate) to the tuning gates, as shown in Fig. 7.17, the occupancy of the left dot will tend to increase in time, while that of the right dot will tend to decrease. Let us suppose that the ramp slopes and starting values are adjusted in such a way that the occupancy of the right dot varies from $M + 1$ to M electrons at the same time as the occupancy of the left dot changes from $N - 1$ electrons to N electrons. Such a situation is represented in the plots of Fig. 7.18: the dashed curve corresponds to the variation in time of the occupancy of the left dot, while the solid curve represents the same quantity for the right dot. The evolution of the applied bias voltages is assumed slow enough to be considered quasi static: this implies that it is slow compared to the time constants and to the tunneling times in the circuit. Parameter values are those used for the first simulation that will be presented in the following. The time dependent behavior of the current I_1 through the bottom pair of dots will be characterized by a peak whenever the occupancies of the two dots vary at the same time, i.e. every time that an electron can be considered as moving between the two dots. Let us now perform the same operation on the upper pair of dots, this time deactivating the voltage sources connected to the lower dots, and using a positive ramp for the right tuning gate (V_7) and a negative ramp for the left gate (V_6) (see Fig. 7.17). It is important to introduce some shift with

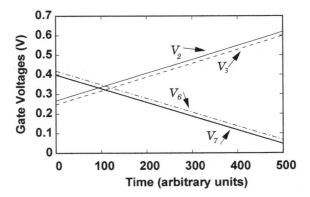

Fig. 7.17.   Voltage ramps applied to the adjustment gates of the QCA cell. Adapted with permission from M. Macucci, M. Gattobigio, and G. Iannaccone, *J. Appl. Phys.* **90**, 6428 (2001). Copyright 2001, American Institute of Physics.

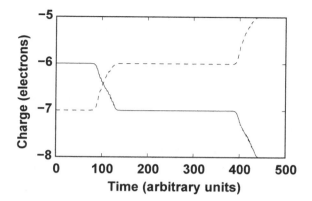

Fig. 7.18. Occupancy as a function of time for the lower left dot (dashed curve) and for the lower right dot (solid curve), when only the lower half of the cell is operated. Adapted with permission from M. Macucci, M. Gattobigio, and G. Iannaccone, *J. Appl. Phys.* **90**, 6428 (2001). Copyright 2001, American Institute of Physics.

respect to the ramps used for the dots in the bottom half of the cell, in such a way that the maxima of I_2 will exhibit a displacement with respect to those of I_1.

Finally, we perform the experiment with both halves of the cell active: the coupling between the upper and the lower half will make the opposite transitions between the upper dots and between the lower dots occur at the same time. Thus the current peaks that would be observed at different times if the two halves of the cell were operated independently, become synchronized, as a result of the electrostatic coupling that is at basis of QCA operation. QCA action can therefore be demonstrated by simply observing the behavior of the two currents I_1 and I_2 as the voltage ramps (with the mentioned shift) are applied to the tuning electrodes: if coupling is sufficient for proper operation, the peaks in the two currents will be simultaneous, otherwise they will occur at different times.

In order to provide quantitative support to our proposal, we have performed numerical simulations with the Monte Carlo technique discussed in the previous chapter. The capacitances of the tunnel junctions, cannot be evaluated from first principles, due to their dependence on the exact lithographic details and on the impurity distribution (which cannot be known exactly), thus we have used values consistent with the experiments: $T_E = 9$ aF and $T = 10$ aF. As far as the geometrical capacitances are concerned, we have instead used the FASTCAP program.[26] Dots are modeled as cylinders with a diameter of 60 nm and a height of 50 nm, while gates are represented with stripes $50 \times 50 \times 250$ nm^3. The interdot distance in the vertical direction is 67.5 nm, while that between the centers of the dots and the edges of the tuning gates is 105 nm. We have chosen a very small interdot distance in order to achieve a value for C_A of the same order of magnitude as that for the tunneling junctions. As far as the relative permittivity is concerned, the one of silicon oxide (4) was adopted, since the silicon nanostructures are embedded in a relatively thick layer of silicon oxide.

A FASTCAP model with a sufficiently large of independent surface panels (600 for the leads and 900 for the dots) has been generated, and a simplified representation (with a smaller number of panels) is provided in Fig. 7.19. The following results were obtained for the relevant geometrical capacitances: $C_D = 0.92$ aF, $C_V = 1.7$ aF, $C_B = 1.2$ aF, $C_C = 0.95$ aF, $C_A = 9.8$ aF.

The capacitances C_K between a gate and the corresponding dot in the opposite pair as well as the the capacitance C_J between a gate and the diagonally opposite dot have also been computed: $C_K = 0.25$ aF, $C_J = 0.21$ aF. A resistance of 5 MΩ has been considered for the tunneling junctions.

Considering the relatively large capacitance value of the tunneling junctions, which leads to a small charging energy, initial simulations have been run for a temperature of just 0.3 K. In the first part of the simulated experiment $V_5 = V_6 = V_7 = V_8 = 0$ to deactivate the upper half of the cell, and voltage ramps with proper slopes (Fig. 7.17) are applied to the adjustment

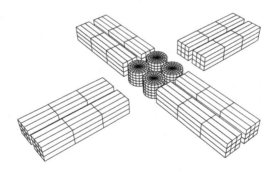

Fig. 7.19. Geometric representation of the cell used for the evaluation of geometrical capacitances (the actual number of discretization panels used for the FASTCAP calculation is much larger, with about 600 panels for the leads and 900 panels for the dots. Reprinted with permission from M. Macucci, M. Gattobigio, and G. Iannaccone, *J. Appl. Phys.* **90**, 6428 (2001). Copyright 2001, American Institute of Physics.

gates of the lower half. The voltage sources supplying the lower double dot, V_1 and V_4, are set to 1 mV. The current I_1 that flows as a result is reported in Fig. 7.20(a) with a dashed line. Then the upper half of the cell is operated while the lower half is disabled, i.e. $V_1 = V_2 = V_3 = V_4 = 0$, $V_5 = V_8 = 1$ mV, and voltage ramps are applied to the upper gates (V_6 and V_7) with the same slopes as V_3 and V_2, with the addition of a 20 mV shift. Due to the shift, the peaks for I_2 (solid line in Fig. 7.20(a)) are displaced with respect to those for I_1.

In the final part of the simulated experiment both halves are operated at the same time: the currents are shown in Fig. 7.20(b), where a clear "locking" effect is visible, with the peaks becoming synchronized between the upper and the lower pair of dots. This locking effect is the result of electrons moving (one at a time) between the two upper dots and forcing,

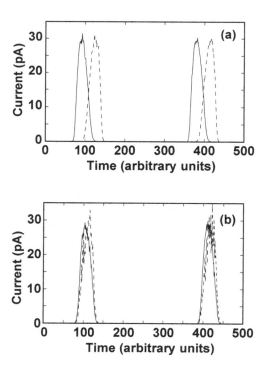

Fig. 7.20. Currents through the two halves of the cell if they are operated one at a time (a) and simultaneously (b), at a temperature of 0.3 K. Adapted with permission from M. Macucci, M. Gattobigio, and G. Iannaccone, *J. Appl. Phys.* **90**, 6428 (2001). Copyright 2001, American Institute of Physics.

through the electrostatic coupling, the opposite transition between the two lower dots. The locking effect is thus equivalent to QCA action (the same correlated electron transfer occurs). For the sake of completeness, we point out that in this context time is just a convenient parameter, while the actual quantities determining the current values are the gate voltages: such quantities have a linear time dependence in the proposed experiment, which allows us to use time as a common parameter.

The achievement of peak locking (and, equivalently, of QCA operation) is strongly dependent on circuit parameters, especially capacitance values: the capacitances C_A create the electrostatic coupling between the upper and the lower dots and must be at least of the same order of magnitude as the other relevant capacitances, among which the tunneling capacitances provide the largest contribution. To obtain the clear locking effect shown in Fig. 7.20 we had to choose a distance between the closest dot edges of just 7.5 nm (for a center-to-center separation of 67.5 nm), which poses very serious fabrication challenges. Even this coupling may become insufficient if the temperature is raised to 4.2 K: in Fig. 7.21 we show the results of a simulation for the same set of parameters considered so far, but with the temperature raised to 4.2 K. It is apparent that the strong thermal broadening of the peaks makes the results for the two cell halves operating at the same time (Fig. 7.21(b)) substantially equivalent to that of the two halves treated separately (Fig. 7.21(a)). If we reduce the C_A capacitances, locking is lost even at 0.3 K, as shown in Fig. 7.22, where the C_A capacitors have been reduced down to 1 aF (this can be obtained simply by increasing the distance between the upper and the lower portions of the cell), a value which is about an order of magnitude smaller than the capacitances of the tunneling junctions, and which destroys QCA functionality. Calculations performed so far were based on the hypothesis of a perfectly symmetric cell; real cells are not symmetric, due to fabrication imperfections, therefore it is important to verify whether the effects that we have discussed so far can be achieved in real-life devices. In the presence of an asymmetry, the periodicity of the current peaks is lost, but we can still obtain the locking effect, although this time, clearly, only on a single peak.

The worst-case asymmetry consists in altering the values of the C_V capacitors: in such a case, the action of the voltage ramps on the corresponding dots becomes unbalanced and the periodicity in the peaks is lost. Ramps can be adjusted in such a way as to obtain the proper shift for a specific peak (the other peaks will behave differently, due to the lost periodicity). Calculations have been performed for the case in which the value

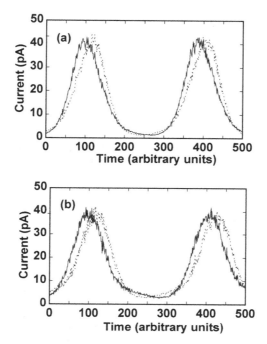

Fig. 7.21. Currents through the two halves of the cell if they are operated one at a time (a) and simultaneously (b), at a temperature of 4.2 K. Adapted with permission from M. Macucci, M. Gattobigio, and G. Iannaccone, *J. Appl. Phys.* **90**, 6428 (2001). Copyright 2001, American Institute of Physics.

of the capacitor C_V between the voltage source V_2 and dot D_1 has been raised to 2.7 aF. Results are reported in Fig. 7.23: the currents in (a) have been computed for the two sections operated independently, while those in (b) have been obtained for a cell with both halves active. It is apparent that locking between the two left current peaks appears, due to the electrostatic coupling between the two halves of the cell. This completes the demonstration of the feasibility of proving QCA cell operation in SOI technology without the need for fabricating charge detectors, which would greatly increase the layout complexity. In particular, we show that, as long as proper bias voltages are applied, geometrical asymmetries can be compensated and do not represent a major problem, if we are only interested in a proof of principle.

Measurements concerning the interaction between both double dots are of great interest for the application of such structures to the implementa-

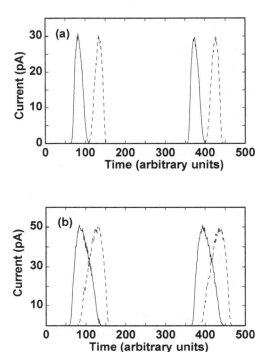

Fig. 7.22. Currents through the two halves of the cell if they are operated one at a time (a) and simultaneously (b), at 0.3 K and with the coupling capacitance C_A reduced down to 1 aF. Adapted with permission from M. Macucci, M. Gattobigio, and G. Iannaccone, *J. Appl. Phys.* **90**, 6428 (2001). Copyright 2001, American Institute of Physics.

tion of QCA cells . We have therefore measured the current through both double dots as a function of the voltage applied to the bottom adjustment gates. The result is shown in Fig. 7.24 as a gray scale plot: for the bottom double dot, the characteristic charging diagram with the honey-comb pattern is obtained, as already discussed, while the current through the top double dot, shown in Fig. 7.24(a), exhibits two distinct features. There are broad current peaks with a large period, which are Coulomb blockade oscillations caused by the small capacitive coupling between the top double dot and the bottom gate. Moreover, a change in conductivity correlated to the transport in the bottom double dot can be observed: each time a current peak in the bottom charging diagram is crossed, a change in the current through the top double dot occurs. This effect is demonstrated more clearly in Fig. 7.25, where a line scan through the charging diagrams of the top

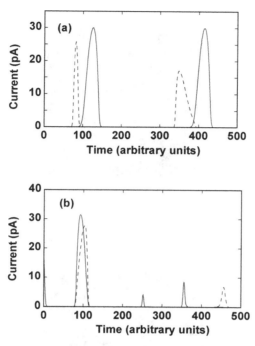

Fig. 7.23.   Currents through the two halves of the cell if they are operated one at a time (a) and simultaneously (b), at 0.3 K and with asymmetric C_V capacitors. Adapted with permission from M. Macucci, M. Gattobigio, and G. Iannaccone, *J. Appl. Phys.* **90**, 6428 (2001). Copyright 2001, American Institute of Physics.

and the bottom double dot is shown for a constant voltage $V_2 = -0.78$ V. The current through the top double dot is indicated with a dashed line, while that through the bottom double dot is represented with a solid line. While crossing a peak in the bottom current, the number of electrons in the bottom structure is changed. This variation in dot charging causes a change in the potential of the top double dot, which in turn results in a change of its conductivity.

The change of the potential seen by the top double dot corresponds to that due to a shift in the side gate voltage V_3 of 30 mV, as indicated in Fig. 7.25, for the peaks around -0.5 V. This is a clear consequence of the electrostatic coupling between the top and the bottom double dots. Using this particular structure, an interdot electron transition in the top double dot forced by an interdot electron transition in the bottom double dot, as it would be required for demonstrating QCA switching, could not

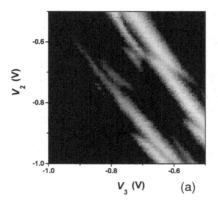

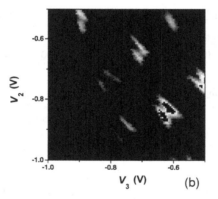

Fig. 7.24. Measured charging diagrams, consisting in a gray scale plot of the source-drain current for the upper (a) and lower (b) double dot as a function of the voltages applied to the two bottom adjustment gates. Bright regions correspond to the maximum current values while black regions indicate zero current.

be observed at this temperature, mainly because of thermal broadening and of insufficient capacitive coupling between the upper and lower dots. The measurements, however, provided the values of the actual capacitances in the structure. In the next section, these values are taken as input for simulations which allow a better understanding of the interaction between the dots and provide leads for further improvement of the layout.

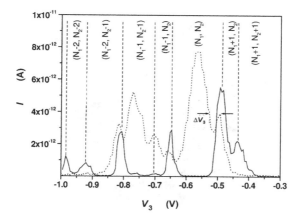

Fig. 7.25. Line scans extracted from the measured charging diagrams, showing the source-drain current through the upper (dashed line) and the lower (solid line) double dots. Arrows mark the shift in the upper double dot current pattern due to tunneling of an electron between the two lower dots.

7.7. Simulations

For the numerical simulations we used the Monte Carlo code presented in Chapter 6. The capacitance matrix used in the simulations is obtained from the measurements. Currents through the top as well as through the bottom double dots were calculated as a function of the side gate voltages V_3 and V_2. Results are shown in Fig. 7.26 and can be directly compared with the experimental data in Fig. 7.24. The period of the oscillations for the bottom double dot as well as the characteristic honey-comb structures are in very good agreement with the measurements in Fig. 7.24, thereby demonstrating the validity of the circuit model of Fig. 7.16.

The arbitrary offset in both voltage axes can be attributed to the arbitrary choice of reference within the periodic honey-comb structure. It is to be noted that, while such a periodicity extends indefinitely in the simulation results, it is limited only to a finite region of the V_3, V_2 plane in actual measurement results, because only in such a region the conditions for correspondence between the actual dots and the lithographically defined ones are satisfied.[30] For the current through the top double dot, the large period Coulomb oscillations are again clearly resolved. The striking features superimposed on these large oscillations are apparent also in the simulations reported in Fig. 7.26.

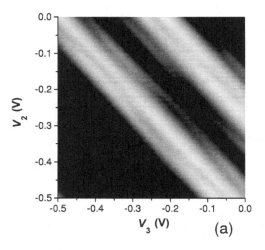

(a)

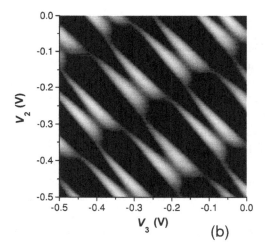

(b)

Fig. 7.26. Simulated charging diagrams, consisting in a gray scale plot of the source-drain current for the upper (a) and lower (b) double dot as a function of the voltages applied to the two bottom adjustment gates. Bright regions correspond to the maximum current values while black regions indicate zero current.

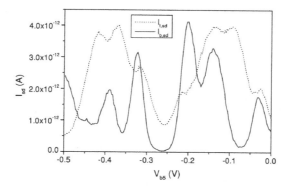

Fig. 7.27. Line scans extracted from the simulated charging diagrams, showing the source-drain current through the upper (dashed line) and the lower (solid line) double dots.

For the sake of a quantitative comparison between simulations and experiments, line scans have been extracted from the simulations: Fig. 7.27 shows line scans for a constant voltage $V_2 = -0.125$ V for both the top and the bottom double dot. As in the measurements reported in Fig. 7.25, each time an extra electron is added to the bottom double dot, the spectrum for the top double dot is shifted along the voltage axis. The size of the shift, e.g. around $V_3 = -0.32$ V, is 32.5 mV, which is in rather good agreement with the measured 30 mV in Fig. 7.25. Similar line scans at other values of V_3 yielded analogous results. From the detailed agreement between measurements and simulations we conclude that a correct interpretation of the electrostatic interaction within the structure has been used. Therefore this simulation approach can be extended to measurements with different configurations of applied voltages, as well as to other structures, providing reliable predictions for the electrical behavior of the devices.

7.8. Possible Improvements

With the information gained in this study, a new candidate structure can be designed, in which coupling between the dots is optimized for QCA applications. From the simulation results it is apparent that to obtain a QCA effect the coupling capacitances between the upper and lower dots need to be of the same order of magnitude as those of the tunneling junctions. There are several possible approaches for achieving this result, which will be discussed below.

A first approach could consist in drastically reducing the capacitances of the tunneling junctions: if it were possible to scale them down by an order of magnitude, the present coupling between the upper and lower dots would already be sufficient to synchronize electron transitions in the upper and lower sections of the cell, with the added benefit of an increase in the operating temperature, due to the enhanced charging energy. Such a reduction of the tunneling junction capacitances has, however, not been investigated in detail, so far, and may involve serious technological difficulties.

Another way of getting capacitances of the same order of magnitude would be that of creating tunneling junctions also between the upper and lower dots. However, tunneling junctions require, as we have seen in the previous sections, a careful tuning of the ratio of the side gate voltages to that of the back gate, in order to actually achieve a single tunneling barrier. New electrodes would be needed to have a sufficient number of degrees of freedom to control six tunneling junctions at the same time, which would greatly complicate the operation of the cell.

An approach leading to an increase of the coupling capacitances without replacing them with tunneling junctions would be represented by linear scaling-down of the device: this would however push lithography to its limits. An asymmetric cell design would partially alleviate demands on lithography. A reduction of the distance between the top and the bottom double dot, while keeping the original distance of 100 nm between the dot centers within each double dot, would yield the desired increase of the capacitances that couple upper and lower dots, such as C_{t1b1}. The cross capacitance between the opposite corner dots, e.g. C_{t1b2}, would be less affected and demonstration of switching for a QCA cell would therefore be easier to accomplish, at least at temperatures in the 0.3 K range.

References

1. T. A. Fulton and G. J. Dolan, *Phys. Rev. Lett.* **59**, 109 (1987); L. J. Geerligs, V. F. Anderegg, P. A. M. Holweg, J. E. Mooij, H. Pothier, D. Esteve, C. Urbina, and M. H. Devoret, *Phys. Rev. Lett.* **64**, 2691 (1990).
2. O. Madelung, *Semiconductor - Basic data*, Springer, Berlin (1996).
3. H. B. Michelson, *IBM J. Res. Dev.* **22**, 72 (1978).
4. C. Kittel, *Introduction to solid state physics*, John Wiley & Sons, New York (1986); P.Y. Yu and M. Cardona, *Fundamentals of Semiconductors*, Springer, Berlin (1996).
5. C. Lebreton, C. Vieu, A. Pépin, M. Mejias, F. Carcenac, Y. Jin and H. Launois, *Microelectron. Eng.* **41/42**, 507 (1998); T. Sato, D. G. Hasko and H. Ahmed, *J. Vac. Sci. Technol. B* **15**, 45 (1997).

6. L. J. Geerligs, V. F. Anderegg, C. A. van der Jeugd, J. Romijn, and J. E. Mooij, *Europhys. Lett.* **10**, 79 (1989).
7. R. Benz, Ph. D. Thesis, University of Tübingen, 1995.
8. E. Schabowska and J. Szczeklik, *Thin Solid Films*, **75**, 177 (1981).
9. K. Matsumoto, M. Ishii, K. Segawa, Y. Oka, B. J. Vartanian and J. S. Harris, *Appl. Phys. Lett.* **68**, 34 (1996).
10. U. Meirav, M. A. Kastner and S. J. Wind, *Phys. Rev. Lett.* **65**, 771 (1990).
11. T. Heinzel, D. A. Wharam, J. P. Kotthaus, G. Böhm, W. Klein, G. Tränkleand, G. Weimann, *Phys. Rev. B* **50**, 15113 (1994); A. T. Johnson. L. P. Kouwenhoven, W. de Jong, N. C. van der Vaart, C. J. P. M. Harmans, and C. T. Foxon, *Phys. Rev. Lett.* **69**, 1592 (1992); J. Weis, R. J. Haug, K. v. Klitzing, and K. Ploog, *Phys. Rev. Lett.* **71**, 4019 (1993).
12. E. Leobandung, L.-J. Guo, Y. Wang, and S. Y. Chou, *Appl. Phys. Lett.* **67**, 938 (1995).
13. H. van Houten, C. W. J. Beenakker, B. J. van Wees, and J. E. Mooij, *Surface Science* **196**, 144 (1988).
14. A. Menschig, A. Forchel, B. Roos, R. Germann, W. Heuring, K. Pressel, D. Grützmacher, *Appl. Phys. Lett.* **57**, 1757 (1990).
15. J.-P. Colinge, *Silicon-on-Insulator Technology: Materials to VLSI*, Kluwer Academic Publishers, Boston (1997).
16. D. Ali and H. Ahmed, *Appl. Phys. Lett.* **64**, 2119 (1994).
17. A. Fujiwara, Y. Takhashi, K. Murase, M. Tabe, *Appl. Phys. Lett.* **67**, 2957 (1995).
18. R. A. Smith, and H. Ahmed, *J. Appl. Phys.* **81**, 2699 (1997).
19. K. Nakazato, and H. Ahmed, *Appl. Phys. Lett.* **66**, 3170 (1995).
20. C. H. Crouch, C. Livermore, R. M. Westervelt, K. L. Campman, A. C. Gossard, *Appl. Phys. Lett.* **71**, 817 (1997).
21. I. M. Ruzin, V. Chandrasekhar, E. I. Levin, L. I. Glazman, *Phys. Rev. B* **45**, 13469 (1992).
22. R. H. Blick, R. J. Haug, J. Weis, D. Pfannkuche, K. von Klitzing, and K. Eberl, *Phys. Rev. B* **53**, 7899 (1996).
23. G. L. Snider, A. O. Orlov, I. Amlani, X. Zuo, G. H. Bernstein, C. S. Lent, J. L. Merz, and W. Porod, *J. Appl. Phys.* **85**, 4283 (1999).
24. F. Hofmann, T. Heinzel, D.A. Wharam, J.P. Kotthaus, G. Böhm, W. Klein, G. Tränkle, G. Weimann, *Phys. Rev. B* **51**, 13872 (1995).
25. M. Macucci, M. Gattobigio, G. Iannaccone, *J. Appl. Phys.* **90**, 6428, (2001).
26. K. Nabors, J. White, *IEEE Trans. Comp. Aided Design Integr. Circ. Sys.* **10**, 1447 (1991).
27. C. Single, R. Augke, F.E. Prins, D.P. Kern, *Semicond. Sci. Technol.* **14**, 1165 (1999).
28. R. Augke, W. Eberhardt, C. Single, F.E. Prins, D.A. Wharam, D.P. Kern, *Appl. Phys. Lett.* **76**, 2065 (2000).
29. R. Augke, W. Eberhardt, S. Strähle, F.E. Prins, D.P. Kern, *Microelectron. Eng.* **46**, 141 (1999).
30. C. Single, F. E. Prins and D. P. Kern, *Appl. Phys. Lett.* **78**, 1421 (2001).

CHAPTER 8

Implementation of QCA Cells in GaAs Technology

Yong Jin

CNRS/LPN
Laboratoire de Photonique et de Nanostructures
Route de Nozay, F-91460 Marcoussis, France

Charles G. Smith

Department of Physics
Madingley Road, CB3 0HE Cambridge, UK

Joan Martorell

Dept. d'Estructura i Constituents de la Materia
Universitat de Barcelona, E-08028 Barcelona, Spain

Donald W. L. Sprung and P. A. Machado

Department of Physics and Astronomy, McMaster University
Ontario L8S 4M1, Hamilton, Canada

Michele Girlanda[a], Michele Governale[b], Giuseppe Iannaccone,
and Massimo Macucci

Dipartimento di Ingegneria dell'Informazione
Università di Pisa
Via Caruso 16, I-56122 Pisa, Italy

---

[a]Current address: Dipartimento di Chimica e Chimica Industriale, Universita' di Pisa, Via Risorgimento, 35, I-56122 Pisa, Italy
[b]Current address: Institut für Theoretische Physik III, Ruhr-Universität Bochum, D-44780 Bochum, Germany

8.1. Introduction

We have also pursued demonstration of QCA operation in the GaAs/AlGaAs material system, which, with respect to SOI, has the disadvantage of a permittivity higher than that of silicon oxide, leading to lower electrostatic interaction, but benefits from the larger body of experience developed for the fabrication of nanostructures defined by means of electrostatic depletion of a 2DEG. In particular, the possibility of finele tuning the tunneling barriers defined by quantum point contacts affords an additional degree of control. Geometric imperfections can be compensated for by tuning the voltages applied to the gates of a layout such as the one discussed in Chapter 3.

The main challenge consists in obtaining a large enough electrostatic coupling: the interaction between the two halves of a cell is limited by how closely the quantum dots can be made. Since the 2DEG is a few tens of nanometers below the surface and therefore distant from the depletion gates, the minimum feature that can be defined in the confinement potential is significantly larger than the minimum achievable lithographic feature. Nevertheless, as detailed in Chapter 9, it has been possible to derive a proof of principle by performing a series of measurements on a GaAs four-dot cell.

In this chapter, instead, we will discuss the technological process used to define QCA cells in the GaAs/GaAlAs material system, the achievable precision in the device geometry, the techniques used for quantitatively reliable modeling, and investigate the decay of a quantum dot that is out of equilibrium with its external environment.

8.2. Nanofabrication of GaAs Devices

In the past two decades, various methods for nanofabrication have been developed and employed widely in academic and industrial laboratories. In principle, the different approaches can be divided into two main categories:[1] direct pattern-generating systems (see Fig. 8.1(a)) and replicating systems (see Fig. 8.1(b)). In the first category, the most versatile technique is electron beam lithography (EBL) which, from a software design, creates a physical pattern in the resist spun on a wafer. A minimum electron beam spot size of less than 10 nm is achieved in most EBL systems. In this way, dimensions of several tens of nanometers or less are accessible. As far as the second category is concerned, the usual approach is photolithography, which transfers a pattern through a photomask into the resist. The "revolution" in this technique in recent years has been the introduction of sophisticated

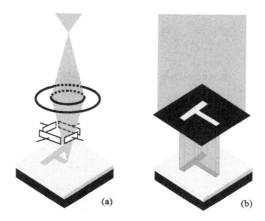

Fig. 8.1. Two lithographic systems: direct pattern generation by means of, e.g. (a) electron beam lithography (EBL), and (b) replication of patterns by photolithography through a photomask.

phase shift masks (PSM). The PSM can reduce the diffraction effect and allows one to reach dimensions below the wavelength of the source light. Currently, features below 100 nm can be defined in this way with very large throughput, suitable for industrial production.

As far as research laboratories are concerned, the greater flexibility of EBL, along with the need for only a few prototype devices, makes it the most convenient choice.

The main fabrication steps in e-beam lithography, using a PMMA (Poly-methyl methacrylate) resist, are illustrated in Fig. 8.2. First, the PMMA resist is deposited on the wafer by spin coating (Fig. 8.2(a)) and baked at 170° C with a hot plate for at least 15 minutes. Then the sample is introduced into an EBL system for patterning by direct e-beam writing with a dose of about $400\mu C/cm^2$, for GaAs wafers. A higher dose of about $600\mu C/cm^2$ is necessary for silicon. The electron beam causes fragmentation of the polymer chains (Fig. 8.2(b)). After exposure, the sample is developed in a solvent consisting in MIBK (Methyl Iso-Butyl Ketone) and IPA (Iso Propyl Alcohol or 2-propanol) with a ratio of 1 : 3 for about one minute, and the exposed portion of the resist is dissolved. The sample is rinsed in IPA, dried with N_2, after which the patterns are available in the resist (Fig. 8.2(c)). It should be mentioned that in practice a combination

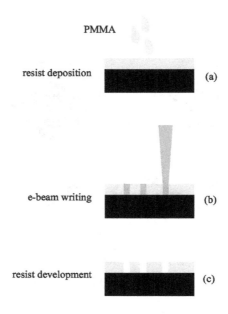

Fig. 8.2.   Main steps for pattern creation with EBL: (a) PMMA resist deposition by spin coating, (b) after baking, the resist is exposed to an electron beam, and (c) the resist is then developed in the solvent composed of MIBK:IPA, and patterns are obtained.

of optical lithography and EBL becomes attractive if the larger patterns of the devices (such as contact pads) are always the same, since a basic photolithography system can be used to create such patterns with a critical dimension of about one micron.

There are two types of pattern transfer through the patterned resist. The first is material deposition, e.g. metals can be deposited on the wafer by evaporation (Fig. 8.3(a)), any metal on the resist can be removed by lift-off (Fig. 8.3(b)) in trichloroethylene solvent. This step is used for forming ohmic contacts or Schottky contacts. The second pattern transfer technique is material removal, the uncovered (by the resist) part of the wafer can be etched with a chemical solution or with RIE (Reactive Ion Etching) (Fig. 8.4(a)), and resist removal (Fig. 8.4(b)). This step is usually applied for mesa definition or gate recess.

Devices with Quantum Point Contacts (QPCs) are based on a 2DEG (two dimensional electron gas) defined by modulation doping in an Al-

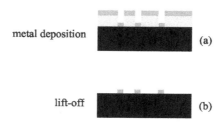

Fig. 8.3. Pattern transfer by material deposition, showing (a) metal evaporation on the patterned resist, and (b) lift-off.

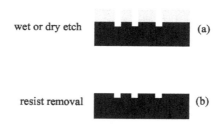

Fig. 8.4. Pattern transfer by material removal, via chemical or dry etching, to (a) remove the material not covered by the patterned resist, followed by (b) resist removal.

GaAs/GaAs heterostructure, which is grown by MBE (Molecule Beam Epitaxy). A typical heterostructure, such as those used in many of our experiments, consists of an undoped GaAs layer, a 20 nm AlGaAs spacer layer, a delta doping layer of Si with a sheet concentration of about 6×10^{12} cm^{-2}, an undoped 10 nm AlGaAs layer, and finally an undoped 5 nm GaAs cap layer. The total distance between the 2DEG and the GaAs surface is only 35 nm. At liquid helium temperature, this 2DEG has an electron mobility of 7.9×10^5 cm^2/(Vs) and an electron density of 4.9×10^{11} cm^{-2}. These high values of mobility and density, and the reduced distance between the 2DEG and the GaAs surface, allow us to realize split gates with a very narrow gap, down to about 35 nm.

The fabrication process with EBL consists of the following four steps. First, a mesa is created, to define the active 2DEG area in the Al-GaAs/GaAs heterostructure: this is done by chemical etching in a solution of H_2O_2:H_3PO_4:H_2O: (1:3:40) at a rate of about 100 nm/min. Then the source and the drain are obtained by creating ohmic contacts. The eutectic alloy is composed by successive metal evaporations of Ni(10 nm)/Ge(60 nm)/Au(120 nm)/Ni(20 nm)/Au(200 nm), followed by a Rapid Thermal Annealing (RTA) at 450 °C for about one minute.[2] Third, a nanometer scale split gate with a gap of about 35 nm is defined by EBL, and a Schottky contact is created with Ti(15 nm)/Au(45 nm). Finally, contact pads and connecting lines are made of Ti(20 nm)/Au(200 nm) (see Fig. 8.5).

Devices with four coupled quantum dots are fabricated by a combination of photolithography and electron beam lithography (EBL), as shown in Fig. 8.6. The device includes 12 shorter electrode pads for ohmic contacts and 12 longer electrode pads for the connection to Schottky gates (Fig. 8.6(a)), obtained by optical lithography. The nanostructured gates defining the four coupled quantum dots (Fig. 8.6(c)) and their connections (Fig. 8.6(b)) to the electrode pads are obtained by EBL. Each quantum

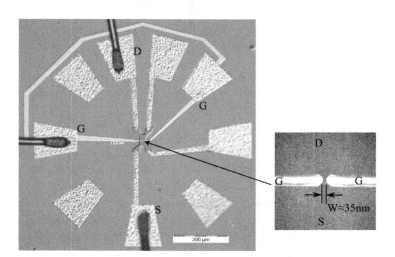

Fig. 8.5.   Overview of the QPC device with the source "S," the drain "D" and the gates "G". The two parts of the split gate are linked by a lithographically defined metal line. The wire bonding is used for electric characterization through the electrode pads "S," "D," and "G". The gap width is about 35 nm.

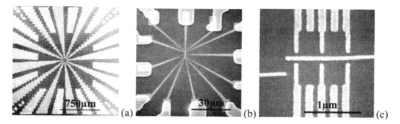

Fig. 8.6. Device with four coupled quantum dots fabricated by a combination of optical and electron beam lithography: the 12 shorter electrode pads used to reach the ohmic contacts, and the 12 longer electrode pads contacting the Schottky gates are obtained by optical lithography (a). The nanostructured gates defining the dots (c), as well as their connections (b) to the electrode pads, are defined by electron beam lithography.

dot is defined by three gates (besides the common horizontal middle gate) with a transverse dimension of about 60 nm. The middle gate is used to control the dot occupancy, while the other two gates are used to adjust the tunneling barriers defining the dots.

As discussed in Chapter 3, the operation of a QCA cell implemented by means of holes in a depletion gate is very sensitive to errors in the diameter of the holes. In order to verify how the best achievable fabrication accuracy compares with that actually needed for the proper operation of a QCA cell without adjustment, we have fabricated an array of nominally identical holes by state-of-the-art electron beam lithography, and then evaluated the symmetry of the cells defined by each group of four holes.[3]

This work was focused in particular on an evaluation of the dispersion of ground state energies due to irregularities in the boundaries of the gate apertures defining both single dots and groups of 4 adjacent dots located at the corners of a square. Two samples were fabricated, one with 60 holes and another with 40 holes: the two samples differ in the average areas of the holes: 5961 nm^2 (with a standard deviation of 349 nm^2) in the first case and 6520 nm^2 (with a standard deviation of 295 nm^2) for the second. This difference is due to the different linewidth used for the lithography of the two samples. An SEM photograph of the first array is shown in Fig. 8.7.

8.3. Evaluation of the Achievable Precision

The first process step consisted in spinning the sample to deposit a 100 nm thick layer of PMMA resist, followed by baking at 170 °C. Exposure was performed in a modified Philips CM20FEG STEM (Scanning Transmission

Electron Microscope) system, which can achieve a spot size of less than 1 nm. The actual linewidth obtained on the resist was varied between 10 and 25 nm, by adjusting the dose between 3 and 6 nCcm^{-2}. Following the exposure phase, development was carried out in MIBK-isopropyl alcohol, and two layers, one of titanium 10 nm thick and the other of gold 20 nm thick were evaporated over the resist. Finally a lift-off procedure was performed with trichloroethylene solvent.

Starting from the SEM image reported in Fig. 8.7 the contour of each dot has been extracted manually, tracking the perimeter of the holes by means of the "xfig" public domain graphic package. The ASCII file produced by xfig, containing the coordinates of the points selected around the hole boundaries has then been processed with dedicated software, capable of computing the area of each hole and the confinement potential that such a hole would produce at a depth of 35 nm from the surface if it were isolated (i.e. if there were just one hole in an otherwise solid gate) or if it were part of a cell, together with three neighboring dots forming a square. To this purpose, a semianalytical procedure outlined in Chapter 3, implying the approximation of Fermi level pinning at the exposed surface, has been applied to the polygons obtained from the graphic extraction procedure. The energy eigenvalues, and in particular the ground state energy, have been obtained by solving the single-particle 2-dimensional Schrödinger equation in the confinement potential resulting from a single dot or from a 4-dot cell.

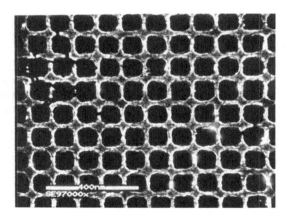

Fig. 8.7.   Scanning electron microscope image of the first sample fabricated for the investigation of fabrication tolerances. Reprinted with permission from M. Macucci, G. Iannaccone, C. Vieu, H. Launois, Y. Jin, *Superlattices and Microstructures* **27**, 359 (2000). Copyright 2000, Academic Press.

The correlation between the computed ground state energies and the dot area has been investigated: results are shown in Fig. 8.8, which is a scatter plot of the ground state energy vs. the reciprocal of the area. A linear fit yields a correlation coefficient of 0.998 for isolated dots and 0.927 for dots included in a cell. The value very close to 1 of the correlation coefficient for single dots shows that the ground state energy is substantially dependent on the hole area, rather than on the detailed shape of the perimeter of the opening. When complete four dot cells are considered, the correlation coefficient goes down because the dimensions of the holes defining the other three dots undergo independent fluctuations which influence the value of the ground state energy: each hole contributes to the definition of the confinement potential of all dots in a cell, although the main contribution is to the dot located directly underneath.

The same scattering plot is reported for the second sample in Fig. 8.9: in this case results are obtained by acquiring the perimeter of each opening independently, and then assuming that they are located at the corners of a 100 nm square (empty dots), at the corners of a 200 nm square (solid dots) or isolated (empty squares). Confirming the previous interpretation, we observe that the correlation between inverse area and ground state energy decreases as the interaction between dots is increased: maximum correlation

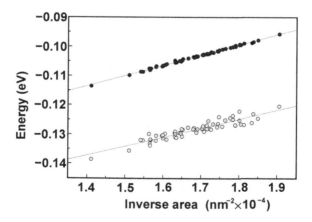

Fig. 8.8. Scattering plot of the ground state energy as a function of the the reciprocal of the area for the sample with average dot area of 5961 nm^2; solid dots represent the data for isolated holes, and the empty dots correspond to four-hole arrangements. Reprinted with permission from M. Macucci, G. Iannaccone, C. Vieu, H. Launois, Y. Jin, *Superlattices and Microstructures* **27**, 359 (2000). Copyright 2000, Academic Press.

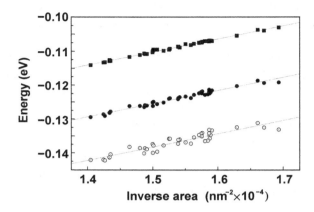

Fig. 8.9.   Scattering plot of the ground state energy as a function of the reciprocal of the area for the sample with average dot area of 6520 nm^2; empty dots represent the data for cells with 100 nm separation, solid dots for 200 nm separation and empty squares for isolated dots. Reprinted with permission from M. Macucci, G. Iannaccone, C. Vieu, H. Launois, Y. Jin, *Superlattices and Microstructures* **27**, 359 (2000). Copyright 2000, Academic Press.

is achieved with isolated dots, while decreasing correlations are obtained as the holes are put closer together.

The data collected from our samples lead to a standard deviation of approximately 4 meV for the ground state energy, corresponding to about 3% of its average value. For a QCA cell to operate properly, the fluctuations in the dot confinement energy must be at least an order of magnitude smaller than the energy separation between the ground state of the cell and its first excited state. For a cell with a separation between dots of 100 nm, a 2DEG depth of 35 nm, a separation from the driver cell of 150 nm (which is already quite small), the difference between the first excited state and the ground state is 0.022 meV; if we increase the intercell separation to more realistic values it is even smaller. Thus the maximum fluctuation in dot energies that we could tolerate for the fabrication of a QCA circuit without external adjustment is of the order of 0.0022 meV, which is about three orders of magnitude below the currently achievable values. We also have to consider that the 4 meV standard deviation that we have computed does not include the effects of randomly located charged impurities nor of any other imperfection that may occur at the interface between the metal gates and the semiconductor. Therefore it is unreasonable to expect that even with improvements in the state of the art a simple "hole array" QCA circuit could be fabricated.

8.4. Electrical Characterization of QPCs

A simplified expression for the quantized conductance G through QPCs, valid for a hard-wall model and neglecting reflections, is [4]

$$G = \frac{2e^2}{h} \frac{W}{\lambda_F/2}, \tag{1}$$

in which λ_F is the Fermi wavelength and W is the one-dimensional channel width. This expression is valid if the mean free path $l_e \gg W$ and $\lambda_F \leq W$. The first condition is expected to be satisfied in our samples, since $l_e = 9.1$ μm for the 2DEG in the heterostructure we have grown. Because of the second condition, quantum effects are expected to become apparent when the split gap in QPCs (see Fig. 8.5) is comparable to the Fermi wavelength λ_F, which is 36 nm for our devices. We have observed proper conductance quantization in DC measurements, as shown in Fig. 8.10, where the current I_{ds} through the QPC for a constant drain-source bias V_{ds} of 1 mV, and the corresponding conductance $G = I_{ds}/V_{ds}$ are plotted as a function of the split gate bias V_{gs}. Quantized conductance can be distinguished up to about 25 K. Conductance quantization is already observed for $V_{GS} = 0$, because depletion underneath the gates occurs as a result of the potential due to the Schottky barrier; in this condition the conductance is approximately $2 \times (2e^2/h)$, which corresponds, based on Eq. (1), to a 1D channel width $W \approx \lambda_F \approx 36$ nm, close to the lithographic gap.

8.5. Modeling of Quantum Point Contacts: The Issue of Boundary Conditions

The choice of boundary conditions for solution of the Poisson equation at the exposed surface of a semiconductor is a thorny issue that has been addressed repeatedly in the literature, without reaching a final answer. Important contributions were given by Larkin and Davies,[5] who provided insight into the consequences of choosing Dirichlet or Neumann boundary conditions, and by Chen and Porod,[6] who devised an interesting procedure involving a distribution of surface states, whose occupancy is determined by solving the Poisson equation in the air above the semiconductor. In particular, the computed pinch-off values for quantum point contacts are extremely sensitive to the choice of surface boundary conditions, and therefore a simulation code for mesoscopic devices based on depletion gates must contain a

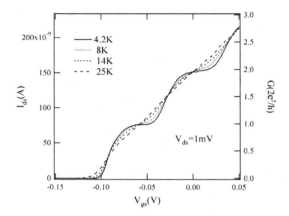

Fig. 8.10. Electric characteristics of the QPC device with a split gap of about 35 nm shown previously. The drain-source current I_{ds} is plotted as a function of the gate bias V_{gs} under a constant drain bias $V_{ds} = 1$ mV. The corresponding conductance $G = I_{ds}/V_{ds}$ is reported in the same figure (with the scale on the right), in units of the conductance quantum $2e^2/h$. Conductance quantization can be observed up to 25 K.

refined treatment of the exposed surface, in order to produce quantitatively reliable results.

With simple Dirichlet boundary conditions the voltage bias needed to pinch-off a quantum dot contact becomes larger than that actually seen in experiments, because of the surface screening effect: a Dirichlet boundary condition at the surface is equivalent to having a conducting plane there, which will significantly reduce the depleting action of the gates in the middle of the quantum point contact. With Neumann boundary conditions, instead, pinch-off is likely to occur at voltages that are (in modulus) lower than the experimental values, because no screening at all from the surface is included.

Within QUADRANT, we have developed a model[7] that will be described in the following, which is able to reproduce correctly the pinch-off voltage for quantum point contacts with not too wide gaps (up to about 100 nm). For wider gaps a more refined model has been developed later[8] and validated in comparison with experimental results.

The method introduced in Ref. [7] is based on a few simple physical concepts: the main problem with the Fermi level pinning concept is that, as bias voltages are varied, charge should flow to the surface to compensate for the new charge distribution inside the device. This is not likely to happen very effectively at cryogenic temperatures, at which charge is instead supposed to be "frozen"[9] and able to move only over large time scales. Based on this observation, we initially study the device with all gates grounded (i.e. with zero bias voltages applied), assuming Fermi level pinning at the surface (we can imagine that the device is cooled down with no bias voltages applied). The difference between the conduction band and the energy at which the Fermi level is pinned is computed by fitting the measured 2-dimensional charge density in the 2DEG in equilibrium conditions. By solving the Poisson equation, we then compute the electric field at the semiconductor-air interface, and, therefore, the surface charge density. Our next step consists in considering such a charge "frozen" in place at the surface, also in the presence of bias voltages, which are applied when the device is already at cryogenic temperatures. Furthermore, due to the step-like variation of permittivity between the semiconductor and the air (12.9 for the semiconductor and 1 for air), we assume that the electric field in the air is negligible and, as a consequence of Coulomb's theorem, keeping constant the surface charge is equivalent to keeping constant the electric field inside the semiconductor: we thus have a Neumann boundary condition for a new solution of the Poisson equation. In this way, the domain outside the semiconductor can be ignored in the solution of the Poisson equation.

Test structures for the validation of our model were fabricated on an AlGaAs/GaAs heterostructure consisting in the following layers. Starting from the bottom: undoped GaAs substrate, 20 nm-thick undoped AlGaAs, silicon delta doping with a dose of 6×10^{12} cm^{-2}, 10 nm-thick undoped AlGaAs, 5 nm GaAs cap layer. The 2DEG forms at a depth of 35 nm, with an electron sheet density $n_s = 4.9 \times 10^{11}$ cm^{-2} at 4.2 K.

Several linear gates with widths ranging from 80 nm to 540 nm have been fabricated as references, in order to extract information useful for the simulation of the split gates. In particular we have used data for the pinch-off voltage of the widest gate, assuming that it can be considered equivalent to an infinite gate. Three types of split gate samples have then been prepared, for comparison with numerical calculations (Fig. 8.11). They have a width of 120 nm and lithographic gaps of 60 nm (E1), 120 nm (F1) and 200 nm (G1). It was necessary, in order to guarantee a good adhesion of the gates to the surface, to perform a 3-4 nm etching before metal evaporation: this affects the pinch-off value for the linear gates and has to be taken into account in the estimation of the fitting parameters.

The first step consists of solving the nonlinear Poisson equation with a 1-D solver, to find the value of the electron density as a function of the Fermi level pinning energy E_{pin} (the difference between the conductance

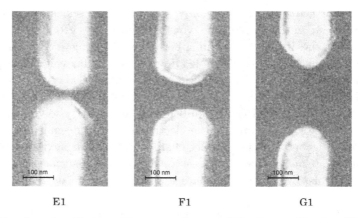

E1   F1   G1

Fig. 8.11. Scanning Electron Microscope pictures of the three split gates: E1 (60 nm gap), F1 (120 nm gap), G1 (200 nm gap). Reprinted with permission from G. Iannaccone, M. Macucci, E. Amirante, Y. Jin, H. Launois and C. Vieu, *Superlattices and Microstructures* **27**, 369 (2000). Copyright 2000, Academic Press.

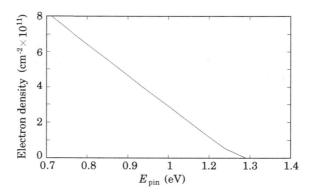

Fig. 8.12. Electron density in the 2-dimensional electron gas as a function of the Fermi level pinning energy. Adapted with permission from G. Iannaccone, M. Macucci, E. Amirante, Y. Jin, H. Launois and C. Vieu, *Superlattices and Microstructures* **27**, 369 (2000). Copyright 2000, Academic Press.

band edge and the energy at which pinning occurs). Results are shown in the plot of Fig. 8.12.

The value of E_{pin}, which leads to an electron density fitting the one experimentally measured, is 0.855 eV, that is within the range proposed in the literature.[10] The corresponding electric field at the surface, inside the semiconductor, is $\mathcal{E} = 0.772$ MV/cm.

The following step consisted of calculating the electron density in the 2DEG as a function of the gate voltage for different choices of the etching depth (Fig. 8.13). There is a shift of 70 mV in the pinch-off voltage for each nanometer of etching. The pinch-off voltages measured on the 540 nm linear gate are 60 mV on one sample and 173 mV on the other, which can be fitted with etching depths of 4 nm and 6 nm, respectively.

The final step consisted of solving the 3D nonlinear Poisson equation for the three samples with split gates, using the actual shape of the gates obtained from the SEM image. The etching depth was assumed to be 4 nm for all cases and the Drude conductance was computed, in the approximation of considering a local conductivity $\sigma(\mathbf{r}) = q\mu n(\mathbf{r})$, where q is the electron charge, μ the mobility, and n the electron density.

Results for the conductance of the E1 and F1 samples are reported in Fig. 8.14(a): the computed pinch-off voltages are -0.22 V and -0.46 V, which compare reasonably well with the experimentally measured values of -0.23 V and -0.39 V. The conductance vs. gate voltage curve for the split gate G1 (which has a 200 nm gap) is reported in Fig. 8.14(b): no pinch-off can be obtained, while the experimental pinch-off value is -0.97 V. This is to be attributed to the already mentioned inadequacy of the surface model for treating split gates with a wider gap. Such gaps can be effectively treated by the more complex model presented in Ref. [8]. The large variation that appears in the measured pinch-off voltage for gate G1 with respect to that

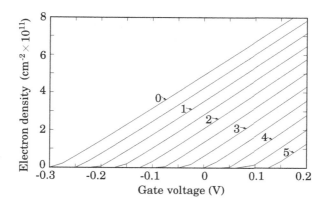

Fig. 8.13.   Electron density in the 2-dimensional electron gas as a function of the gate voltage, for different etching depths (expressed in nanometers). Adapted with permission from G. Iannaccone, M. Macucci, E. Amirante, Y. Jin, H. Launois and C. Vieu, *Superlattices and Microstructures* **27**, 369 (2000). Copyright 2000, Academic Press.

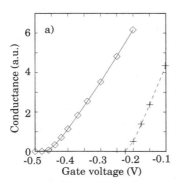 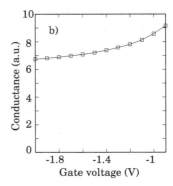

Fig. 8.14.   Computed conductance vs. gate voltage for the split gates E1 and F1 (a), and for the split gate G1 (b). Reprinted with permission from G. Iannaccone, M. Macucci, E. Amirante, Y. Jin, H. Launois and C. Vieu, *Superlattices and Microstructures* **27**, 369 (2000). Copyright 2000, Academic Press.

of gate F1 is also evidence of a qualitative change in the behavior when the gap is increased from 120 nm to 200 nm.

8.6. Electron Decay from an Isolated Quantum Dot

8.6.1. *Lifetimes of the experimentally studied dot*

Electron decay from a well isolated quantum dot was studied by Cooper *et al.*[11] In their experiment, metal gates were fabricated by electron beam lithography on the top surface of an MBE grown GaAs/AlGaAs heterostructure in which a 2DEG layer sat approximately 70 nm. below the surface. A schematic of the gates is shown as an insert in Fig. 8.15. This is one corner of the device shown in Fig. 8.6(c), which has been discussed above. Negative voltages applied to the gates deplete the 2DEG directly beneath, thus defining the dot and detector structure. The measurements were made in a dilution refrigerator with a base temperature of 100 mK. Modulated bias voltages of 50 and 300 μV were applied to the QD and the detector simultaneously, and the output current from each passed to lock-in amplifiers. The dot and detector circuits were kept electrically isolated and the bias voltages were modulated at well separated frequencies.

Initially the system was set up to study Coulomb oscillations as the voltage on one of the gates, called the plunger P, was swept. From this study, the dot was estimated to contain 300 electrons with the gates just at pinch-off. The charging energy was found, by summing the total capacitances of all gates, to be 0.5 meV. The 1D detector was biased to be close to pinch-off, with resistance of around 100 kΩ. It was then sensitive enough to variations in the electric potential of the QD that it effectively acted as an electrometer with resolution of $e/10$, allowing the detection of a change of charge by a single electron in the adjacent QD.

To study the lifetimes, the dot was first filled by reducing the barrier voltages on gates C_1 and C_2 to -0.21 V and the plunger P to -0.2 V at $t = 0$; the gates C_i were then rapidly (within a second) ramped to -0.44 V, and then P to -0.4 V. After that, the voltages were maintained for one hour while the detector signal was monitored. The detector signal showed steps as each electron escaped from the dot, giving a staircase-like trace. The step heights were relatively small compared to the working range of the detector, making its response effectively linear, and giving almost equal step heights over a range of up to ten electrons. Normally, from one hour-long trace to the next, the total drift in the detector signal was less than the step size for a single electron decay. This means that one could compare one trace to the next and apply a normalization to each, which allowed the decay signals of equivalent electrons to be identified.

Escape times were recorded in a series of 154 runs, each of which lasted for one hour, during which up to 10 sequential escapes were recorded. Assuming that each run is an independent experiment on an identical quantum dot, the escape times can be analyzed to provide a set of mean lifetimes for the escape from an identifiable level of the dot.

Some support for this view is provided by the modeling of the system[12] by Martorell *et al.* They were able to reproduce semi-quantitatively the observed lifetimes by self-consistent calculations on the dot properties. The underlying theory is that the dot results from the combined effect of the heterostructure, the applied gate voltages, and the electrostatic repulsion of the confined electrons. After each electron escape, the system must re-equilibrate, and the succeeding escape is from a slightly different configuration of the dot. We suppose this relaxation to occur on a time scale much shorter than the observed escape times, which are in the tens or hundreds of seconds. The situation is analogous to a chain of radioactive decays in nuclear physics, where the daughter nucleus is a distinct species with its own properties and decay lifetime.

What was recorded are the times at which the current in the detector circuit signaled a change in the occupancy of the dot. Those for decay one always exceed 9 s, which takes account of the initial settling down of the dot, during which several short-lived decays may have occurred. We assume that $t_0 = 9$ is the effective start time for the decay processes. By taking differences of these times, one obtains escape times for the successive decays. Since the charging and subsequent counting was repeated 154 times, (of which two recorded no signal,) at most one could have 152 observations of the 10 decays which were seen. Since counting stopped after one hour, it often occurred that the later decays, especially the eighth to tenth, were not seen, resulting in considerably fewer observations. If the eighth decay is observed at time $t^{(8)}$, but not the ninth, then $3600 - t^{(8)}$ is a lower bound on the time at which the ninth decay occurred.

8.6.2. *Statistical analysis of the experimental data*

In the original data file there are some cases for which there was a missing value (represented by NA). Some examples are shown in Table 8.1:

Sometimes, as in run #94, one of the intermediate decays was not seen. This might be because it was simultaneous with the preceding or the following escape. The easiest thing is to omit these cases, or to take the difference of the following and the preceding escape times as an upper limit on the succeeding one: here the fourth decay $t^{(4)} < 143$ s. In run #12, neither the first or second escapes was recorded; this gives an upper limit on $t^{(3)} < 67$ s. For the first seven decays there are more than a hundred valid data available, and while we looked at the consistency of these upper limits, they were not directly taken into account in the lifetime analysis.

Another situation is when, as in run #12, the seventh escape was recorded but none of the subsequent ones. If we assume that the eighth decay did in fact not occur within the hour, then we have a lower limit $t^{(8)} > 2538$ s. Beyond decay six the number of valid data declines rapidly, due to the increase of the lifetime. Then these lower limits play an important role in the analysis.

A constant probability per unit time is assumed for a given decay event, leading to an exponential distribution of decay times with a characteristic

Table 8.1. Some examples of electron escape times from the experiment

decay #	1st	2nd	3rd	4th	5th	6th	7th	8th	9th	10th
run #	...	...	...	...	...	...	...	...	...	...
12	NA	NA	67	223	290	432	962	NA	NA	NA
71	NA	17	42	198	204	230	829	1792	3035	NA
94	17	22	NA	165	170	315	462	NA	NA	NA
...	...	...	...	...	...	...	...	...	...	...
Total=152	Valid=107	107	129	137	148	146	124	72	23	2

lifetime τ:

$$f_\tau(t) = \frac{1}{\tau} e^{-t/\tau}. \tag{2}$$

8.6.3. *First decays*

For the first few decays we simply ignore the invalid readings and for a sequence of $n < 152$ valid events we define the joint probability function (also called maximum likelihood function) as

$$M(\tau) = \prod_{i=1}^{n} f_\tau(t_i) = \frac{1}{\tau^n} \exp\left(-\sum t_i/\tau\right). \tag{3}$$

The most likely lifetime $\hat{\tau} = \sum t_i/n$ maximizes $\ln(M)$. If we suppose that the n different runs are independent, the variable $2\sum t_i/\tau$ obeys the χ^2_{2n} distribution[14] with $2n$ degrees of freedom. Using this distribution we deduce confidence limits on the extracted lifetimes, as shown in Table 8.2:

8.6.4. *Later decays*

For the later decays, only a small fraction of the $n = 152$ runs show decay signals. Suppose for example that on the i-th run, the sixth decay is observed at time $t^{(6)}$, but the hour ran out before the seventh was seen. The assumption we make is that $t^{(7)} \geq 3600 - t^{(6)} = L_i$ gives a lower bound on the decay time of this seventh decay. The maximum likelihood function is then defined as

$$M(\tau) = {}_nC_r \prod_{i=1}^{r} \frac{1}{\tau} e^{-t_i/\tau} \prod_{r+1}^{n} e^{-L_i/\tau} \tag{4}$$

where the factors $e^{-L_i/\tau}$ arise from the probability S that an electron will survive until time L_i without decay (ie. $S = \int_L^\infty e^{-t'/\tau} dt'/\tau$). The τ that maximizes this M is

$$\hat{\tau} = \frac{\sum_1^r t_i + \sum_{r+1}^n L_i}{r}. \tag{5}$$

Table 8.2. Lifetime estimates (90% confidence bands), using only firm data

Decay #	1	2	3	4	5	6	7	8	9	10
data used	107	107	129	137	148	146	124	72	23	2
Upper Estimate	8.0	17.3	43.9	125	214	530	777	1157	1213	3855
Estimate	6.3	14.6	37.8	108	186	460	666	942	829	685
Lower Estimate	5.0	12.6	32.9	94	163	403	577	784	607	289

The situation is similar to what is called type I censorship of data in statistics.[14] There one starts with n identical systems and observes for a fixed time L, during which some number r of the systems decay. Our situation is only a bit more complicated by the fact that the L_i are not identical. For purposes of determining the confidence intervals for $\hat{\tau}$, Cox[13] concluded that for type I censorship the distribution of $(2r\hat{\tau})/\tau$ should obey the χ^2_{2r+1} distribution with $2r + 1$ degrees of freedom, where $\hat{\tau}$ is the most likely estimate. We used this to obtain error bands on $\hat{\tau}$:

$$\text{Prob}\left(\chi^2_{2r+1}(\alpha/2) \leq \frac{2r\hat{\tau}}{\tau} \leq \chi^2_{2r+1}(1 - \alpha/2)\right) = 1 - \alpha \tag{6}$$

where α is the chosen confidence percentage.

Applying this method to the experimental data gives the results summarized in Table 8.3 below, and illustrated in Fig. 8.15.

Table 8.3 contains our best estimates of the mean lifetimes. Compared to Table 8.2, they increase steadily with decay number. For decays 6 to 10 they take account of the lower limits provided by the missing last escapes; this information was omitted in the original analysis.

8.6.5. Modeling of electron decay from the isolated quantum dot

It has become commonplace to say that a quantum dot is an artificial atom, but in fact the self-consistent potential confining electrons in a large dot has more in common with the mean field potential in a heavy nucleus: flat in the interior, with abrupt walls. An artificial nucleus is a more apt description, as will become clear in what follows. Indeed, the detection of sequential decays from an isolated quantum dot is a more favorable situation for study of the decay process, as the question of preforming the alpha particle inside the nucleus is a difficult complication. Hence, we can more confidently test our knowledge of the confining barriers for electrons, as well as the profile, and dependence on occupation number, of the dot potential. These questions were considered in Ref. 12.

Table 8.3. Lifetime estimates (90% confidence bands), with type I censorship

Decay #	1	2	3	4	5	6	7	8	9	10
data used	107	107	129	137	148	150	148	124	72	23
Upper Estimate	8.0	17.2	43.7	125	214	622	1274	2565	4658	18367
Estimate	6.3	14.6	37.8	108	186	543	1111	2208	3822	12884
Lower Estimate	5.0	12.5	32.8	94	163	476	973	1909	3161	9260

In that work we studied the decay process using analytic models based on realistic numerical simulations of the confinement potential. As the dot contains about 300 electrons, Poisson-Thomas-Fermi calculations should be adequate to describe the electron density and the confining potential of the dot. From those we developed accurate analytic approximations for the confining potential that allowed us to construct an envelope approximation wave function for the electrons in the dot, and to compute the electron lifetimes from the transmission amplitude across the barrier.

Previous work on quantum dots was more concerned with the wave functions of confined states in the dot, the electron density distribution and the shape of the confining potential. For such purposes, only the inside of the barrier matters. It is when one looks at the escape of electrons from the dot that the barrier height, its width, and its shape become important; these are crucial to understand the lifetimes.

Electron decay from a strongly isolated quantum dot was studied in Ref. 11. Such results are shown in Fig. 8.15 together with a sketch of the device (inset). In the following we present the experimental procedure and the modeling of the dot decay.

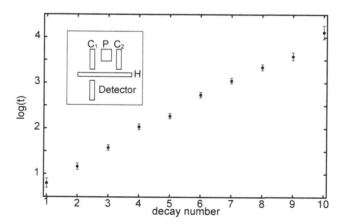

Fig. 8.15. Lifetimes of electrons escaping from a quantum dot. Insert: schematic depiction of the gates defining the dot and the detector. Adapted with permission from J. Martorell, D. W. L. Sprung, P. A. Machado, and C. G. Smith, *Phys. Rev. B* **63**, 045325 (2001). Copyright 2001, American Physical Society.

8.6.6. *Theoretical framework*

The Poisson-Thomas-Fermi modeling was described in Ch. 3, so here we list only the main steps:

i) First, Poisson-Schrödinger (PS) and Poisson-Thomas-Fermi (PTF) simulations as described in Ref. [15] were performed for the ungated heterostructure. The inputs for the PS simulation were the thickness and composition of each layer in the heterostructure, and the dopant concentration in the donor layer. The only adjustable parameter is the donor ionization energy which is set to be $e\Phi_i = 0.12$ eV, in order to reproduce the measured 2DEG density, $n_e = 2.74 \times 10^{11}$ cm^{-2}. For the simpler Poisson-Thomas-Fermi scheme we employed a common relative permittivity $\varepsilon_r = 12.2$ for all layers of the heterostructure, which, combined with the parameters already used for the PS simulation, also reproduce the experimental n_e. After this "fitting" the model has no other free parameters.

ii) For the gated structure we used the gate layout and voltages of the experiment. To solve the Poisson equation for the gated heterostructure one has to impose as a boundary condition the value of the electrostatic potential on the exposed surface of the heterostructure, and on the gates. We assumed Fermi level pinning and chose the energy of the surface states as the zero of the energy scale. In this convention, the conduction band edge is set at $eV_s = 0.67$ eV on the exposed surface. Under each gate the conduction band is set at $eV_{ms} + eV_g$, where V_g is the gate voltage and the metal semiconductor contact potential, eV_{ms}, is taken as 0.81 eV.[17] The electrostatic potential due to the gates is then computed using semi-analytic expressions based on the work of Davies *et al.*[18] Added to this were: *a)* the direct Coulomb potential between the electrons, and a mirror term which imposes the boundary conditions at the surface, and *b)* the contribution from the fully ionized donor layer and its mirror term (see Sec. IIA of Ref. [16] for details of a similar example.) We neglect exchange and correlation effects, which are small.

iii) The connection between the confining potential defined by the conduction band edge and the electron density is completed by using the Thomas-Fermi approximation at zero temperature:

$$\rho_e(\vec{r}) = \frac{1}{3\pi^2}\left(\frac{2m^*}{\hbar^2}(E_F - eV(\vec{r}))\right)^{3/2}. \tag{7}$$

The PTF iteration was performed starting from the ungated heterostructure densities as trial values.

8.6.7. *Equilibrium dot*

We began with the dot in its final state after all excess electrons have escaped. This corresponds to a PTF simulation with the same Fermi level, $E_{F,dot} = 0$, for the electrons in the dot and in the 2DEG outside the barriers. The gate voltages are taken from Ref. [11] as $V_{PL} = -0.40 \ V$, $V_{C1} = V_{C2} = -0.44 \ V$ and $V_H = -0.7 \ V$. The predicted PTF 3D electron distribution $\rho_e(x, y, z)$ is more conveniently visualized in terms of a projected 2D density:

$$n_e(x, y) = \int_{z_j}^{\infty} \rho_e(x, y, z) \ dz \ , \tag{8}$$

where z_j is the junction plane. The $n_e(x, y)$ distribution, shown in Fig. 8.16, has an approximately rectangular boundary, and its maximum value is close to the 2DEG density of the ungated heterostructure. In this calculation the dot contains 286 electrons.

8.6.8. *Dot with excess electrons*

To study these configurations we set the Fermi level inside the dot, $E_{F,dot}$, higher than its value outside the barriers, $E_{F,2DEG} = 0$. We can do so because the dot is well pinched off from the surrounding electron gas. We ran PTF simulations with equally spaced values for $E_{F,dot}$ running from 0

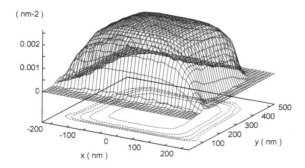

Fig. 8.16.   The two-dimensional PTF density, $n_e(x, y)$, for a dot in equilibrium with the surrounding 2 DEG. Reprinted with permission from J. Martorell, D. W. L. Sprung, P. A. Machado, and C. G. Smith, *Phys. Rev. B* **63**, 045325 (2001). Copyright 2001, American Physical Society.

to 17.5 meV. The occupation Q of the dot increases linearly with $E_{F,dot}$ at the rate 2.75 electrons per meV, giving occupations $286 \leq Q \leq 334$.

The simulations also produced the confining potential for the electrons in the dot, $eV(x, y, z)$. To reduce this to a two-dimensional function, $U(x, y)$, we took a weighted average over the density profile in the z direction:

$$U_{PTF}(x, y) = \frac{\int_{z_j}^{\infty} eV(x, y, z) \ P(z) \ dz}{\int_{z_j}^{\infty} P(z) \ dz} \qquad (9)$$

where

$$P(z) = \int_{\Omega} \rho_e(x, y, z) \ dx \ dy \ . \qquad (10)$$

Here the domain of integration Ω was a rectangle in the x, y plane extending a short distance into the surrounding electron gas, (from $(x_l, y_l) = (-510, -255)$ nm to $(x_r, y_r) = (510, 255)$ nm. This includes an area outside the dot where the 2DEG is still depleted by the gates. Although the computed $V(x, y, z)$ is not separable, previous experience with Poisson-Schrödinger simulations of wires[16,19] and circular dots, has shown us that the factorization ansatz leads to very good approximations when the z degree of freedom is integrated out as in Eq. (9).

In Fig. 8.17 we show the $U_{PTF}(x, y)$ corresponding to the equilibrium dot of Fig. 8.16. As expected from the gate layout shown in the inset to Fig. 8.15, it has two very high barriers running parallel to the x axis, one centered at $y = 0$ and the other that begins with a steep rise at $y \simeq 400$ nm (and shows clearly the mark of the three-fingered gate layout labeled $C1$, $C2$, and PLunger in Fig. 8.15). Tunneling across these barriers is negligible. In addition there is a symmetric pair of barriers running parallel to the y axis, with maxima at $x \simeq \pm 238$ nm through which the electrons *do* tunnel. In the interior, the potential is practically constant. Although these x-barriers slightly increase in height with increasing y, the rectangular shape of the potential suggests using a separable approximation in Cartesian coordinates:

$$U_{PTF}(x, y) \simeq U_s(x, y) = U(x) + W(y) \ . \qquad (11)$$

We will interpret the experimental decay data using this separability ansatz. For the $W(y)$ barriers, which are basically impenetrable, we used two simple models described below. As a guide to a realistic choice for the x-dependent term we examine in Fig. 8.18 the profiles of $U_{PTF}(x, y)$ at a fixed value of $y = 200$ nm in the middle of the dot. The profiles shown

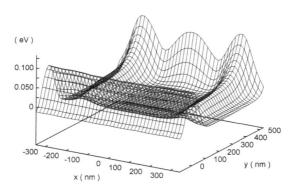

Fig. 8.17. Two-dimensional confining potential, $U_{PTF}(x, y)$ for the dot of Fig. 8.15. Reprinted with permission from J. Martorell, D. W. L. Sprung, P. A. Machado, and C. G. Smith, *Phys. Rev. B* **63**, 045325 (2001). Copyright 2001, American Physical Society.

cover a range of occupations of up to forty excess electrons. In this range, the potential at the dot center increases linearly with Q, according to

$$U_0 = 0.347Q - 118.4 \text{ meV} . \qquad (12)$$

At large distances outside the dot, $U_\infty = -18.8$ meV is constant. Similarly, the location of the barrier maximum and its height can be parametrized as:

$$x_b = 238 - \frac{Q - 286}{16} \text{ nm}$$
$$U_b = 0.117Q - 13.4 \text{ meV} . \qquad (13)$$

Note that $dU_b/dQ \approx (1/3) \, dU_0/dQ$ reflects the decrease of the screened Coulomb repulsion away from the center of the dot. Furthermore, we found the x-dependence to be very well reproduced (see Fig. 8.18) using the following analytic model:

$$U(x) = U_b + U_{MF}(x) , \quad x > 0,$$
$$= U(-x) , \qquad\qquad x < 0, \quad \text{where}$$

$$U_{MF} \equiv U_c \frac{\sinh^2\left(\frac{x - x_b}{w_b}\right)}{\cosh^2\left(\frac{x - x_b}{w_b} - \mu\right)} . \qquad (14)$$

This potential form has the great advantage that the transmission coefficient for U_{MF} is known analytically.[20] U_{MF} is an asymmetric barrier which

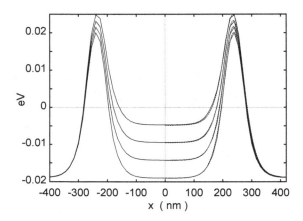

Fig. 8.18.   Continuous lines: sections at $y = 200$ nm of the $U_{PTF}(x, y)$ corresponding to $E_{F,dot} = 0.0, 0.005, 0.010$ and 0.015 eV. Dashed lines: analytic parametrization for $U(x)$ as described in the text (the latter shown only for $x > 0$ for clarity.) Reprinted with permission from J. Martorell, D. W. L. Sprung, P. A. Machado, and C. G. Smith, *Phys. Rev. B* **63**, 045325 (2001). Copyright 2001, American Physical Society.

takes one value for $x << x_b$ and another value for $x >> x_b$:

$$U_{MF}(x_b) = 0$$
$$U_{MF}(\infty) \equiv \lim_{x \to \infty} U_{MF}(x) = U_c e^{2\mu}$$
$$U_{MF}(-\infty) \equiv \lim_{x \to -\infty} U_{MF}(x) = U_c e^{-2\mu} . \tag{15}$$

The parameters U_b, U_c, μ, x_b, w_b allow one to fit the barrier height, the potential floors inside and outside the dot, the barrier spacing and the barrier width. Since the barriers are spread quite far apart, in practice $x_b >> w_b$, so $U_{MF}(x = 0) \approx U_{MF}(-\infty)$. In this case,

$$U_0 \equiv U(0) \approx U_b + U_c e^{-2\mu}$$
$$U_\infty \equiv \lim_{x \to \infty} U(x) = U_b + U_c e^{2\mu}. \tag{16}$$

Then we can solve for

$$\mu = \frac{1}{4} \ln\left(\frac{U_b - U_\infty}{U_b - U_0}\right) \qquad \text{and}$$
$$U_c = -(U_b - U_0)e^{2\mu} . \tag{17}$$

To determine the parameters appearing in Eq. (14), we took the values of the PTF potential at the origin, U_0, well beyond the barrier, U_∞, and the value U_{x_b} at the barrier maximum $x = x_b$, and then plot $U(x)$ to find the best w_b, which turned out to be 48 nm. This gives a convenient analytic form for the confining potential, motivated by PTF, whose transmission coefficient is:

$$T = \frac{2\sinh(\pi k_+)\sinh(\pi k_-)}{\cosh(\pi(k_+ + k_-)) + \cosh(\pi\beta)} \quad , \tag{18}$$

where

$$k_{-/+} = \sqrt{\frac{2m^*}{\hbar^2}(E - U_{0/\infty})w_b^2} \quad \text{and}$$

$$\beta = \sqrt{\frac{2m^*}{\hbar^2}(2U_b - 2U_c - U_0 - U_\infty)w_b^2 - 1} \quad . \tag{19}$$

Barrier shape $W(y)$: In Fig. 8.19 we examine a section of $U_{PTF}(x = 0, y)$ through the center of the dot. We used two trivial models, the simplest one being an infinite square well, of width $w_y \approx 350$ nm. The slightly fancier one is a truncated harmonic oscillator:

$$W_{tho}(y) = \qquad 0 \qquad\qquad \text{(flat bottom)}$$
$$= -0.13 + \frac{1}{2}k_y(y - y_0)^2 \quad \text{(walls) .} \tag{20}$$

with $y_0 = 238$ nm and $k_y = 7.35 \cdot 10^{-6}$ nm^{-2}. As can be seen in Fig. 8.19, this parametrization (plus the constant term U_0) reproduces the main features of the $x = 0$ sections of the PTF potentials.

By combining Eqs. (12–20) we had a separable analytic potential model for the dot containing a desired number Q of electrons. This removes the necessity of repeatedly solving the PTF equations for the self-consistent field, while studying the decay process.

8.6.9. Quasibound states of the dot

We constructed the electron wave functions inside the dot in the envelope function approximation, using our parametrized potential, $U_s(x, y)$. The single electron energies are

$$E_{n_x, n_y} = E_{n_x} + E_{n_y} \tag{21}$$

corresponding to

$$\Psi_{n_x, n_y}(x, y) = \phi_{n_x}(x)\psi_{n_y}(y) . \tag{22}$$

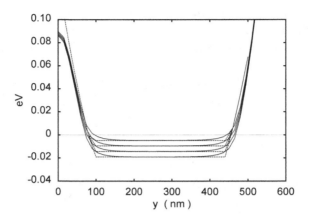

Fig. 8.19.   Continuous lines: sections at $x = 0$ nm of the $U_{PTF}(x, y)$ for the same $E_{F,dot}$ as in Fig. 8.17. Dashed lines: analytic parametrization of $W_{tho}(y) + U_0$ as described in the text. Reprinted with permission from J. Martorell, D. W. L. Sprung, P. A. Machado, and C. G. Smith, *Phys. Rev. B* **63**, 045325 (2001). Copyright 2001, American Physical Society.

The factors satisfy 1D Schrödinger equations:

$$-\frac{\hbar^2}{2m^*}\phi''_{n_x}(x) + U(x)\phi_{n_x}(x) = E_{n_x}\phi_{n_x}(x)$$
$$-\frac{\hbar^2}{2m^*}\psi''_{n_y}(y) + W(y)\psi_{n_y}(y) = E_{n_y}\psi_{n_y}(y)\,. \tag{23}$$

The second equation is for a confined wave function, easily solved by standard numerical methods. We label the solutions by the number of extrema, n_y, of the eigenfunction. For example, taking $W(y)$ to have hard walls, the energy is

$$E_{n_y,sw} = \frac{\hbar^2}{2m^*}\left(\frac{n_y\pi}{w_y}\right)^2\,. \tag{24}$$

For the truncated harmonic oscillator shape there is no similar analytic expression, but the dependence on n_y is similar.

   The x-dependent equation describes 1D electrons confined in the dot by the "leaky barriers". Weakly quasibound state solutions were computed using methods described in Ref. [21]. However, for levels corresponding to the long tunneling lifetimes observed in the experiment, the energies and eigenfunctions can be computed well enough by the simpler prescription of setting the electron wave function to zero at the points $\pm x_b$ inside the barriers. Furthermore, if only the eigenvalues and lifetimes are needed, we have checked that the WKB quantization condition is adequate:

$$\int_{x_l}^{x_r} \sqrt{\frac{2m^*}{\hbar^2}(E(n_x) - U(x))}\ dx = \left(n_x - \frac{1}{2}\right)\pi \, . \tag{25}$$

The lifetime $\tau = 1/\lambda$ is the inverse of the "decay constant", defined as the number of "decays" per second per parent "dot". For a single dot the electron wave function is normalized to unity over the volume inside the barriers, and λ for a given level is just the outgoing flux at large distance. When the decay probability is small, one can treat the electron as confined in the dot. Classically, its trajectory will oscillate between the right, x_r, and left, x_l turning points, with a period

$$P = 2\int_{x_r}^{x_l} \frac{dx}{v(x)} = 2\int_{x_r}^{x_l} \frac{dx}{\sqrt{\frac{2}{m^*}(E_x - U(x))}} \, , \tag{26}$$

The flux λ is then given by the frequency of hits against the barriers, $2/P$, times the transmission probability T across a barrier, and therefore:

$$\tau = \frac{1}{\lambda} = \frac{1}{T}\int_{x_l}^{x_r} \frac{dx}{v} \, . \tag{27}$$

This expression is very convenient because the transmission coefficient Eq. (18) for our parametrized potential, $U(x)$, is known analytically.[20]

To "construct" the desired dot configuration with Q excess electrons, we generated $U_s(x,y)$ for the chosen value of Q, and filled the levels as follows: a) First we list the (E_{n_x}, τ_{n_x}), in order of increasing n_x (and therefore of increasing energy and decreasing lifetime). This list is truncated at an $n_x = n_{x,max}$ whose lifetime is less than 0.01 sec. b) Next we formed a list of 2D levels (n_x, n_y) by choosing those for which

$$E_{n_x} + E_{n_y} \leq E_{n_{x,max}} + E_{n_y=1} \, . \tag{28}$$

The levels in this list are occupied in order of increasing energy and according to Fermi statistics, see Eqs. (29),(30). We chose the dot Fermi level to obtain the desired Q. It is supposed that, for the long lifetimes observed in the experiment, the electrons have time to lose energy by phonon collisions and occupy the quasibound states of lowest energy. Since the dot is located inside a crystal at temperature T', the electrons should be at the same temperature. The level occupations $f(E)$ are determined by Fermi statistics:

$$f(E) = \left[1 + e^{\frac{E - E_F}{k_B T'}}\right]^{-1} \, , \tag{29}$$

where these are now 2D energies. The Fermi level is obtained from

$$Q = \sum_{i=(n_x,n_y)} 2f(E_i), \qquad (30)$$

with 2 for spin degeneracy. For the ensemble of electrons in the dot, the flux λ will now be the sum of fluxes for each occupied single particle level, weighted by the level occupancy:

$$\lambda = \sum_{i=(n_x,n_y)} 2f(E_i)\lambda_i \qquad (31)$$

and the corresponding half-life is again $\tau = 1/\lambda$.

To produce a sequence of decays for comparison to experiment we proceeded as follows: *i)* We start with a dot containing a number of electrons, Q_0, chosen large enough so that the lifetime for one electron to escape is under five seconds. *ii)* We redetermine the barrier and dot configuration for $Q = Q_0 - 1$ electrons, as described in the above paragraph and determined again the corresponding lifetime for escape of one electron. This process is repeated to generate a sequence of decays that covers and extends beyond the range of lifetimes measured in experiment. From that list we chose as the first observed electron decay that corresponding to the Q whose lifetime is the first to be larger than $t_0 = 5$ seconds.

8.6.10. *Results and discussion*

In Fig. 8.20 we show results from our model, using parameters chosen as described above, for a range of lifetimes extending over three orders of magnitude. The stars correspond to the truncated harmonic oscillator choice for $W(y)$, whereas the $+$'s are for the square well choice (with a value $w_y = 380$ nm chosen to optimize the agreement with the other prescription in the range of experimental lifetimes, from 10 to 1000 seconds.) One sees that the trends are very similar. For Q in the neighborhood of 304, the predicted decay lifetimes fall in the experimental range.

As already mentioned, our PTF simulations predict $Q = 286$ for the dot in equilibrium with the surrounding electron gas. This is also what we find with this separable model, as the curve of lifetimes shown in figure 8.20 extrapolates smoothly up to $Q = 287$, for which we found a lifetime of $\log_{10} \tau = 5.2$, or 44 hours. After that, the Fermi level of the electrons inside the dot falls below that of the surrounding 2DEG and further decays are blocked. It should take almost two days for the dot to reach equilibrium with its surroundings.

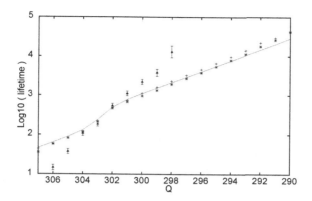

Fig. 8.20. Calculated lifetimes (in seconds) when $W(y)$ is either the truncated harmonic oscillator: stars, or a square well with $w_y = 380$ nm: + signs . The dotted line is the prediction of the two level model, Eq. (33). The experimental points from Fig. 8.15 have been placed arbitrarily. Reprinted with permission from J. Martorell, D. W. L. Sprung, P. A. Machado, and C. G. Smith, *Phys. Rev. B* **63**, 045325 (2001). Copyright 2001, American Physical Society.

For values $Q < 300$, we find a linear dependence of lifetime on charge. Such linear dependence occurs when our model produces a sequence of decays dominated by those from a single 1D electron level; i.e. corresponding to a fixed value of n_x. To understand why, suppose that at zero temperature and for Q electrons, the occupied level with shortest lifetime is $(n_{x,s}, n_y)$, and that $\{n'_x, n'_y\}$ are occupied levels with higher energy and longer lifetime (this requires that at least $n'_x < n_{x,s}$ for longer lifetime and $n'_y > n_y$ for higher total energy). When one forms the $Q - 1$ electron configuration according to the rules explained above, one of the $\{n'_x, n'_y\}$ levels will be empty, whereas the level $(n_{x,s}, n_y)$ will again be filled. In more physical terms: all the electrons with energy above that of the level with shortest lifetime will lose energy by phonon collisions and fall into the leaky level, from which they finally escape. Since the lifetime does not depend on n_y, all the electrons with energy above that of the state $(n_{x,s}, n_y = 1)$ will escape through the same leaky 1D level, $n_{x,s}$, which remains the favored decay channel as long as it is occupied. Therefore the total probability for **one** electron to escape from the occupied states with quantum number $n_{x,s}$ is the probability for a single 1D electron with energy $E_{n_{x,s}}$, multiplied by the number of electrons in occupied states with the same quantum number

$n_{x,s}$: $q_{n_{x,s}}$:

$$\tau(Q) = \frac{\tau_{n_{x,s}}(Q)}{q_{n_{x,s}}(Q)} \quad , \tag{32}$$

and when the occupation $q_{n_{x,s}}$ of the leaky level is constant, the linear variation of $\log_{10}(\tau)$ reflects that of the lifetime of the leaky level. This is where the 2D nature of the quantum dot asserts its presence, even though the decay appears to proceed only in one dimension.

In our calculations for $Q < 300$, the occupation of the $n_x = 13$ level stays practically constant and $n_x = 14$ level remains empty. For higher values of Q both levels contribute significantly to the escape lifetime. In this situation:

$$\tau(Q) = \frac{1}{\dfrac{q_a(Q)}{\tau_a(Q)} + \dfrac{q_b(Q)}{\tau_b(Q)}} \quad . \tag{33}$$

This is shown as the dotted curve in Fig. 8.20, and it accounts very well for the trend of the lifetimes predicted by the separable model.

A separable model favors the appearance of the linear decay sequences, because of the degeneracy in lifetime of states with the same $n_{x,s}$. A non-separable model would lift that degeneracy and then the lifetime sequences should show a behavior intermediate between the two situations discussed above. In particular, the sudden change of slope at $Q = 302$ in Fig. 8.20 would presumably spread over a wider range of values of Q. Not surprisingly, the predicted lifetimes for the observed decays depend sensitively on details of the barrier shape.

In Fig. 8.21 we also compare the calculated lifetimes with the revised experimental ones from Fig. 8.15; the latter are placed on the assumption that the first decay recorded was from the dot with 307 electrons. For decays in the range from 100 to 1000 seconds, qualitative agreement with calculations using the standard set of parameters is apparent. In Fig. 8.21 we varied the barrier thickness within a range of -4% to $+3\%$. This changes the slope of the lifetime vs. charge somewhat, but the experimental lifetimes seem to be increasing faster than we can reproduce. For longer and shorter lifetime decays, the situation is worse. Experimental lifetimes more than double for each successive decay while theory increases at just half that rate. What has changed from Ref. [12] is the lifetimes: we have done a new analysis of the decay sequences and are now confident in the extracted mean lifetimes.

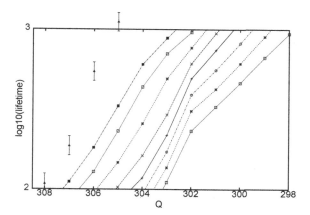

Fig. 8.21.   Slow decays: Calculated lifetime sequences corresponding to variations of the standard barrier width from -3% (right) to $+4\%$ (left) in steps of 1%. The $+$ signs joined by a continuous line (to guide the eye) correspond to the prediction for the standard set of parameters. Experimental points with 90% error bars taken from Fig. 8.15 with the origin of Q shifted arbitrarily. Reprinted with permission from J. Martorell, D. W. L. Sprung, P. A. Machado, and C. G. Smith, *Phys. Rev. B* **63**, 045325 (2001). Copyright 2001, American Physical Society.

The shortest lifetimes are for particles closer to the top of the confining barriers. These may be more sensitive to the approximations made in the theory. In particular, the separability assumed in our work may be the culprit. The barriers for decay in the x-direction are not flat as y increases, so the decays may occur preferentially where the barriers are lower. A proper two-dimensional calculation of the decay process would be required to see how valid the separability approximation is.

References

1. W. B. Glendinnig and J. N. Helbert, *Handbook of VLSI Microlithography*, Noyes Publications (1991); P. Rai-Choudhury, *Handbook of microlithography, micromachining, and microfabrication*, SPIE Press (1997).
2. Y. Jin, *Solid-State Electronics* **34**, 117 (1991).
3. M. Macucci, G. Iannaccone, C. Vieu, H. Launois, Y. Jin, *Superlattices and Microstructures* **27**, 369 (2000).
4. B. J. van Wees, H. van Houten, C. W. J. Beenakker, J. G. Williamson, L. P. Kouwenhoven, D. van der Marel, and C. T. Foxon, *Phys. Rev. Lett.* **60**, 848 (1988).

5. I. A. Larkin and J. H. Davies, *Phys. Rev. B* **52**, R5535 (1995).
6. Minhan Chen and W. Porod, *J. Appl. Phys.* **78**, 1050 (1995).
7. G. Iannaccone, M. Macucci, E. Amirante, Y. Jin, H. Launois and C. Vieu, *Superlattices and Microstructures* **27**, 369 (2000).
8. G. Fiori, G. Iannaccone, M. Macucci, *Journal of Computational Electronics* **1** 39, 2002.
9. S. E. Laux, D. J. Frank, F. Stern *Surface Science* **196**, 101 (1988).
10. S. Sze, *Physics of semiconductor devices*, Wiley, New York (1981), p. 270.
11. J. Cooper, C. G. Smith, D. A. Ritchie, E. H. Linfield, Y. Jin and H. Launois, *Physica E* **6**, 457 (2000).
12. J. Martorell, D. W. L.Sprung, P. A. Machado and C. G. Smith, *Phys. Rev. B* **63**, 045325 (2000).
13. D. R. Cox, *Biometrika* **40**, 354 (1953).
14. J. F. Lawless, *Statistical Models and Methods for Lifetime Data*, Wiley, New York (1982), Ch. 3.
15. J. Martorell and D. W. L. Sprung, *Phys. Rev. B* **49**, 13750 (1994).
16. J. Martorell, Hua Wu and D. W. L. Sprung, *Phys. Rev. B* **50**, 17298 (1994).
17. W. Mönch, *Semiconductor Surfaces and Interfaces*, Springer (1995).
18. J. H. Davies, *Semicond. Sci. Technol.* **3**, 995 (1988); J. H. Davies, A. Larkin and E. V. Sukhorukov, *J. Appl. Phys.* **77**, 4504 (1995).
19. J. Martorell and D. W. L. Sprung, *Phys. Rev. B* **54**, 11386 (1996).
20. P. M. Morse and H. Feshbach, *Methods of Theoretical Physics* McGraw Hill (1953) p. 1651.
21. J. Killingbeck, *J. Phys. A* **10**, L99 (1977); *Phys. Lett. A* **78**, 235 (1980).

CHAPTER 9

Non-Invasive Charge Detectors

Giuseppe Iannaccone, Carlo Ungarelli[a], Michele Governale[b]
and Massimo Macucci

Dipartimento di Ingegneria dell'Informazione
via Caruso 16, I-56122 Pisa, Italy

Spiros Gardelis, Charles G. Smith, John Cooper, David A. Ritchie
and Edmund H. Linfield

Department of Physics
Madingley Road, CB3 0HE Cambridge, UK

Yong Jin

CNRS/LPN
Laboratoire de Photonique et de Nanostructures
Route de Nozay, F-91460 Marcoussis, France

9.1. Introduction

Reading the state of a cell without disrupting the operation of the whole system is a critical requirement for the actual implementation of quantum cellular automata architectures.

A key enabling technology has therefore been the development of non-invasive single charge detectors,[1] which exploit the variation of the resistance of a quantum point contact when a electron is added to (or removed

[a]Current address: Physics Department "Enrico Fermi", University of Pisa, Largo Pontecorvo 3, 56127 Pisa, Italy
[b]Current address: Institut für Theoretische Physik III, Ruhr-Universität Bochum, D-44780 Bochum, Germany

from) a closely placed dot. The electrostatic potential defining the constriction is modified by the contribution of the additional electron, so that the transmission coefficients and, consequently, the overall resistance are affected.

Such single charge detectors have been recently applied to a number of cases, and in particular in systems of multiple dots,[2,3] also for detecting the state of few-electron dots.[4]

9.2. Experiments on a Double Dot System with Non-Invasive Detector

Figure 9.1 shows the gate layout of the system we have fabricated, that allows us to either define four independently controllable dots, i.e. a complete QCA cell, or two coupled dots with quantum constrictions that can be used as detectors.

With the application of a negative bias, all the gates showed the same initial pinch-off voltages at ≈ -0.25 V. However, the pinch–off characteristics of each of the nominally identical gates showed one of two possible

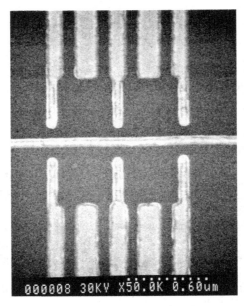

Fig. 9.1.   Gate layout of the system that allows us to define four dots (a QCA cell) or two dots plus quantum constrictions that can be used as detectors.

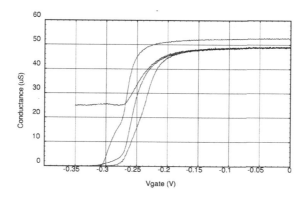

Fig. 9.2. Conductance of quantum point contacts.

behaviors: either a sharp decline to pinch-off (occurring at ≈ -0.3 V, see Fig. 9.2) or else a long tail with the final pinch–off at around -0.7 V. Such a variation in behavior is probably due to the fact that the smallest dimensions of the gate widths and gaps are significantly smaller than the depth of the 2DEG. In principle, such a condition should lead to a sharp cut–off, but surface impurities in close proximity to the gate may produce some localized screening, which in turn causes the long tails.

At the beginning, the system is set as a single large dot with the center gate at a voltage well below pinch off. As the center gate is swept, the single large dot transforms into a dot pair and, as can be seen in Fig. 9.3, the period of the CB oscillations halves. In this trace approximately 200 electrons can be seen leaving the dot, which is estimated to be about one third of the total initially contained within it.

As a second step, we investigated the detection of sealed dots. In order for quantum cellular automata to become a usable technology, it is vital to be able to measure the occupancy of dots that are not coupled to the outside electron gas (2DEG), while most measurements of dot states in the literature have been made using Coulomb Oscillations in the single electron tunneling regime, in which the electrons are strongly coupled to the surrounding electron reservoirs.

Fig. 9.4 shows a the detector signal for a series of plunger gate sweeps made on a single dot system, with the barrier height between the dot and surrounding 2DEG being increased in each trace. In the first (upper) sweep, the dot and 2DEG are strongly coupled and the electrons exit and then re-enter freely as the plunger gate voltage is first increased and then subsequently reduced.

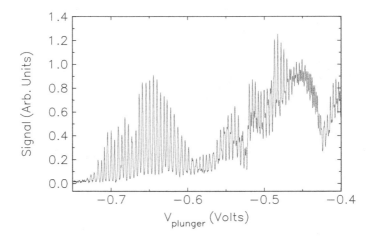

Fig. 9.3. Detector signal of large dot with swept center gate. Note that the period changes as the center gate pinches off and the single large dot turns into a pair of small ones.

In the second trace, the QPC barriers are made higher so as to decrease the coupling, and it can be seen that the "plunger-in" and "plunger-out" traces are no longer identical, indicating that the expelled electrons cannot easily re-enter the dot. As an analogy, the system may be thought of as a beaker being lifted clear out of a bath of water, as the beaker is squashed the water it contains will be squeezed out and falls into the bath below, but if it is returned to its original size the water will not go back in.

By the third trace the "plunger-in" and "plunger-out" curves are quite different. While it is still possible to see electrons being forced out, there is no sign of any re-entering, and the parabolic curve of the "plunger-out" line reflects the direct unscreened coupling of the plunger gate on the detector.

9.3. Numerical Simulation of the Dot-Detector System

In order to investigate theoretically the behavior of the non-invasive voltage probe, and to determine the criteria for its optimization, we address the simulation of a system consisting of a single quantum dot plus detector, electrostatically defined by metal gates evaporated on an AlGaAs-GaAs

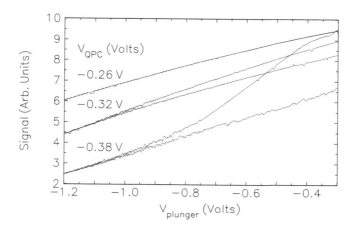

Fig. 9.4. Detector Signal as the plunger gate voltage of a single dot is increased and decreased. As V_{QPC} is increased the dot becomes increasingly isolated.

heterostructure identical to that on which the experimental devices have been fabricated.[5]

In particular, we consider the following AlGaAs-GaAs heterostructure, sketched in Fig. 9.5: an undoped GaAs substrate, an undoped AlGaAs spacer layer of 20 nm, a Silicon delta doping layer of 6×10^{12} cm^{-2}, an undoped AlGaAs layer of 10 nm, an undoped GaAs cap layer of 5 nm. An unintentional acceptor doping of 10^{15} cm^{-2} is considered in all undoped layers.

The quantum dot and the nearby constriction are electrostatically defined by the voltages applied to the top aluminum gates, whose layout is shown in Fig. 9.6: gates numbered as 1, 2, 3 and 4 define the quantum dot, with a geometrical area of 188×104 nm; the constriction between gates 4 and 5 forms the detector.

In order to evaluate the sensitivity of the non-invasive detector, we compute dot occupancy and the resistance of the quantum wire as a function of the voltage applied to the plunger gate (gate 2) for several wire resistances. The bias voltages V_i of gates i ($i = 1 \dots 4$) are -0.12 V, while we have considered different voltages V_5 for gate 5, between -0.15 and 0.2 V, corresponding to initial resistances of the quantum constriction ranging from 7.6 KΩ to practically infinity.

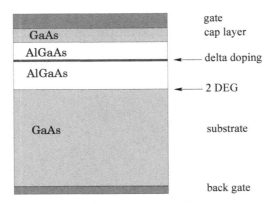

Fig. 9.5. Schematic view of the AlGaAs-GaAs heterostructure on which the experimental dot-detector systems have been fabricated and considered in the simulations. The layer structure consists of a GaAs substrate, 20 nm undoped AlGaAs spacer, Si delta doping layer (6×10^{12} cm^{-2}), 10 nm undoped AlGaAs, 5 nm GaAs cap layer. Reprinted with permission from G. Iannaccone, C. Ungarelli, M. Macucci, E. Amirante, M. Governale, *Thin Solid Films* **336**, 145 (1998). Copyright 1998, Elsevier Science S. A.

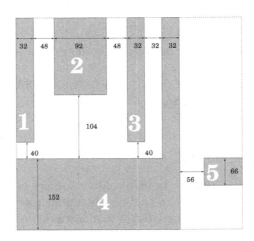

Fig. 9.6. Gate configuration defining the quantum dot and the detector (dimensions are in nm). Gate 2 is the plunger gate; the voltage applied to gate 5 modulates the resistance of the detector. Reprinted with permission from G. Iannaccone, C. Ungarelli, M. Macucci, E. Amirante, M. Governale, *Thin Solid Films* **336**, 145 (1998). Copyright 1998, Elsevier Science S. A.

In order to assess the functionality of the detector, the voltage of the plunger gate can be swept toward more negative values, so as to progressively deplete the quantum dot, and the induced quantum wire resistance must be calculated. The whole simulation, including the solution of the 3-D Schrödinger-Poisson problem and a recursive Green's function calculation[7] of the conductance in the thus computed potential profile should be therefore repeated a large number of times, and would be prohibitively time consuming.

For this reason, we have chosen a less rigorous approach, which is, however, much simpler from the computational point of view. In detail, we start by solving the nonlinear Poisson equation on a 3D grid ($65 \times 65 \times 65$ points) with a semiclassical approximation:

$$\nabla \cdot (\epsilon \nabla \phi) = -q(p - n + N_D^+), \tag{1}$$

where ϕ is the electrostatic potential, ϵ is the dielectric constant, q is the electron charge. The semiclassical hole and electron concentration (p and n, respectively) are obtained as[6]

$$p = N_V \frac{2}{\sqrt{\pi}} F_{1/2} \left(\frac{E_V - E_F}{kT} \right) \tag{2}$$

$$n = N_C \frac{2}{\sqrt{\pi}} F_{1/2} \left(\frac{E_F - E_C}{kT} \right), \tag{3}$$

where N_C and N_V are the effective density of states in the conduction and in the valence band, respectively, E_C and E_V are the conduction and valence band edges, respectively, k is the Boltzmann constant and T the absolute temperature. $F_{1/2}$ is the Fermi-Dirac integral

$$F_{1/2}(y) = \frac{1}{1/2!} \int_0^\infty \frac{x^{1/2}}{1 + e^{x-y}} dx. \tag{4}$$

N_D^+ is the concentration of ionized donors, given by

$$N_D^+ = \frac{N_D}{1 + g_D \exp \frac{(E_F - E_D)}{kT}}, \tag{5}$$

where N_D is the donor concentration, g_D the spin degeneracy factor (2 for GaAs), E_D the donor energy level. The band edges depend on the potential through

$$E_C(\mathbf{r}) = E_{00} - \chi(\mathbf{r}) - \phi(\mathbf{r}), \tag{6}$$

where E_{00} is the vacuum energy level of the electrode assumed as a reference for the potential ϕ and the voltage, χ is the electron affinity. In addition $E_V(\mathbf{r}) = E_C(\mathbf{r}) - E_g(\mathbf{r})$, where E_g is the energy gap.

As far as the boundary conditions are concerned, Dirichlet conditions are enforced for the potential at the gate surfaces (i.e. $\phi = E_{00}/q - V_i - \phi_m^{(i)}$, where V_i is the voltage applied to gate i and $\phi_m^{(i)}$ is its work function) while on the lateral boundary regions of the simulation domain, Neumann boundary conditions with zero perpendicular electric field are enforced.

The choice of proper boundary conditions at the exposed GaAs surface requires some attention, and has been performed on the basis of the "frozen charge" model already discussed in Chapter 8.

As far as the Fermi level at equilibrium is concerned, we have chosen the value which provides the best fit of the pinch-off voltage with our experiments, i.e. 5.25 eV below the vacuum level, corresponding to a normal component of the electric field at the surface of 88.2 V/μm.

We should mention the fact that we have subsequently developed a much more accurate method for taking into account states at the semiconductor-air interface, that has been validated through specific experiments on purposely fabricated test structures.[8]

At this point the contribution to the potential from the charge in the dot is calculated by solving again the Poisson equation, using the charge in the dot volume Ω_{dot} as the only source term, i.e.

$$\nabla \cdot (\epsilon \nabla \phi_{\text{dot}}) = +q n_{\text{dot}}, \tag{7}$$

where $n_{\text{dot}} = n$ for $\mathbf{r} \in \Omega_{\text{dot}}$, $n_{\text{dot}} = 0$ otherwise. The confining potential for the electrons in the dot is therefore

$$V_{\text{conf}} = E_C + q \phi_{\text{dot}}. \tag{8}$$

The many-body Schrödinger equation should be solved with the confining potential given by (8) on a 3D domain. However, we can assume that confinement in the vertical direction is much stronger than that on the horizontal plane, so that only the first vertical subband is occupied, in order to decouple the 3D problem into a 2D problem at the heterointerface plane, and a 1D problem in the vertical direction.

In particular, we solve the single particle Schrödinger equation in the vertical direction (along the z-axis):

$$-\frac{\hbar^2}{2m_z} \frac{\partial^2 \chi(z)}{\partial z^2} + V_{\text{conf}}(x_0, y_0, z)\chi(z) = E\chi(z), \tag{9}$$

where the pair (x_0, y_0) defines a point on the horizontal plane. Let the ground state eigenvalue and eigenfunction be E_0 and $\chi_0(z)$, respectively: the confining potential seen by electrons at the heterointerface plane is the

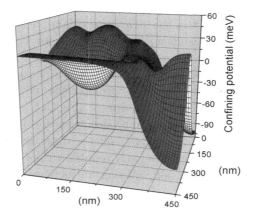

Fig. 9.7. Confining potential for the electrons in the 2DEG at the GaAs-AlGaAs heterointerface. Reprinted with permission from G. Iannaccone, C. Ungarelli, M. Macucci, E. Amirante, M. Governale, *Thin Solid Films* **336**, 145 (1998). Copyright 1998, Elsevier Science S. A.

given by $V_{2D}(x, y) = V_{conf}(x, y, z_0) + E_0$ where z_0 is the z-coordinate of the 2DEG. In Fig. 9.7, the V_{2D} is shown for the case of $V_5 = -0.16$ V.

The many-body 2D Schrödinger equation is then solved within the framework of density functional theory, and reads:

$$-\frac{\hbar^2}{2m_x}\frac{\partial^2\varphi_i}{\partial x^2} - \frac{\hbar^2}{2m_y}\frac{\partial^2\varphi_i}{\partial y^2} + [V_{2D} + V_{Hartree} + \qquad (10)$$

$$V_{ex} + V_{corr}]\,\varphi_i = \epsilon_i\varphi_i, \qquad (11)$$

where $V_{Hartree}$ is he Hartree term for the Coulomb interaction, V_{ex} and V_{corr} are the exchange and correlation terms given by theory of Tanatar and Ceperley.[9] The details of the numerical model can be found in Macucci *et al.*[10] The electron-electron interaction is modeled consistently with the Neumann conditions enforced at the exposed surface, i.e. with negative image charges, which warrant constant electric field perpendicular to the surface. The number of electrons in the dot is that which gives the highest possible chemical potential below the Fermi level in the bulk.

Finally, the confining potential in the quantum wire is calculated by adding to the previously obtained confining potential the contribution of the electron density in the dot computed with the Schrödinger-Poisson solver.

When the plunger gate voltage is modified, instead of solving again the 3D Poisson equation, we use a semianalytical method[11] to evaluate the

correction to the confining potential on the plane of the 2DEG, assuming
that the other charges in the structure remain unchanged.

In Fig. 9.8 the results of the simulation for $V_5 = -0.16$ are shown:
for a plunger gate voltage V_2 of -0.61 V the dot is completely depleted
and the detector resistance is 12.9 kΩ. As V_2 is raised in steps of 10 mV,
the confining potential on the heterointerface plane is lowered, therefore
the number of electrons in the dot N (Fig. 9.8 - bottom) progressively
increases. The detector resistance decreases for increasing plunger gate
voltage as long as N is constant; on the other hand, when one electron
is added to the dot, the Coulomb repulsion rises the confining potential
of the quantum constriction, causing an increase of a few percent of the
detector resistance.

We have also studied the dependence of the detector sensitivity upon
the initial resistance of the quantum constriction. In Fig. 9.9 the detector
resistance is plotted as a function of the plunger gate voltage for four different voltages applied to gate 5: of course, the lower the voltage applied to
gate 5, the higher the initial detector resistance. As can be seen, a high sensitivity can be obtained if conduction in the quantum wire is essentially in

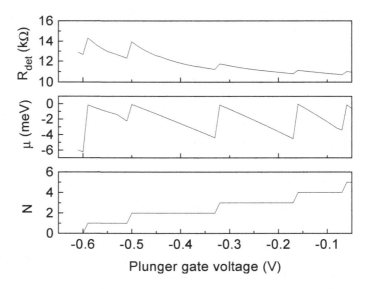

Fig. 9.8. Computed detector resistance (top), electrochemical potential (center) and
number of electrons in the dot (bottom) as a function of the voltage applied to the
plunger gate.

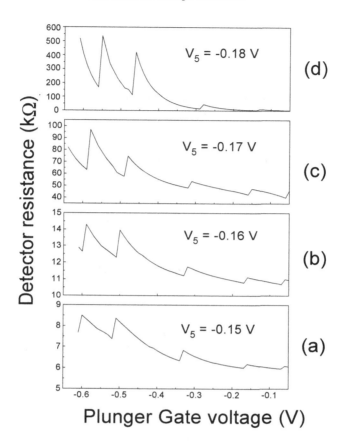

Fig. 9.9. Detector resistance as a function of the voltage applied on the plunger gate for four different values of the voltage V_5. For V_5 lower the -0.18 V, the quantum wire has negligible conductance. Reprinted with permission from G. Iannaccone, C. Ungarelli, M. Macucci, E. Amirante, M. Governale, *Thin Solid Films* **336**, 145 (1998). Copyright 1998, Elsevier Science S. A.

the tunneling regime, as in the cases of Figs. 9.9 (d) and (c), corresponding to an initial resistance much higher than 12.728 kΩ, i.e. that associated to a single propagating mode in the quantum wire. This is simply due to the fact that the transmission probability in the case of tunneling is extremely sensitive to a variation of the confining potential profile. For values of V_5 lower then -0.18 V the quantum wire is practically pinched off.

Therefore, simulations confirm that the electrostatic coupling between the dot and the constriction seems to be a viable detection principle for quantum cellular automata systems. A more complete simulation, including the four dots and the associated detectors, would be useful to really verify whether the "stray" capacitive couplings between detectors and between each detector and the different dots can undermine correct detection.

9.4. Determining the Operation of a AlGaAs-GaAs QCA Cell

Here, we briefly review recent experiments showing an experimental evidence of the correct operation of the QCA cell defined by the structure shown in Fig. 9.1. We will mainly refer to the paper by S. Gardelis et al.[2] where the interested reader can find all details.

In particular, the experiment has shown that by moving an electron between the two bottom dots of Fig. 9.1, we can trigger an electron switching between the two top dots, in the opposite direction, so that we obtain switching of the polarization of the whole cell.

Let us refer to the gate layout illustrated in the inset of Fig. 9.10. The principle of the experiment is very simple. First, gates are biased in order to form the two dots C and D, and the constrictions defined by gates G_4 and G_6 are used as a detector. In addition, the voltages on gates G_7 and G_{11} are sufficiently low that dots C and D are almost completely isolated from the rest of the 2DEG. This means that electrons can only move from C to D and viceversa, but the sum of the electrons on dots C and D is constant.

Then, the voltage on gate G_{10} ($V_{G_{10}}$) is swept toward more negative values, in order to move electrons from dot C to dot D. First the curve of the detector conductance G as a function of $V_{G_{10}}$ is decomposed into a smooth component $G_{average}$ and a high frequency component ΔG ($G = G_{average} + \Delta G$). Then, the detector signal, plotted in Fig. 9.10(a), is obtained as $\Delta G/(dG_{average}/dV_{G_{10}})$.

Let us consider the variation of the detector signal as $V_{G_{10}}$ is swept toward lower voltages. On average, the conductance decreases, since when $V_{G_{10}}$ is lowered the confining potential for electrons in the constriction is raised. Almost periodically, we observe that an electron tunneling from dot C to dot D is energetically favored, and this occurs causing an abrupt variation of the detector signal (the sign of such a variation depends on the detailed electrostatics of the device).

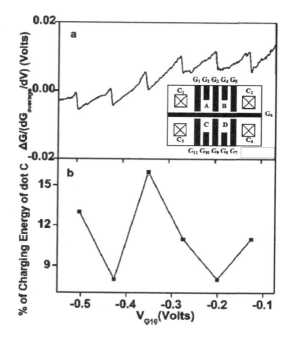

Fig. 9.10. Inset: gate layout considered in the experiment; a) Detector signal as a function of the voltage applied to gate G_{10}; b) ratio of the energy required for switching an electron between dots C and D to the charging energy of dot C. Adapted with permission from S. Gardelis, C. G. Smith, J. Cooper, D. A. Ritchie, E. H. Linfield, Y. Jin, *Phys. Rev. B* **67**, 033302 (2003). Copyright 2003, American Physical Society.

If we assume that dots C and D are almost identical, and that the variation of $V_{G_{10}}$ has a much smaller effect on the chemical potential of dot D with respect to dot C, we can interpret the distance between two peaks in Fig. 9.10(a) as proportional to twice the Coulomb charging energy of dot C. From the same figure, the voltage shift required for the abrupt change of ΔG to take place is proportional (with the same constant) to the energy required for switching the polarization of the pair of dots C, D.

As can be seen in Fig. 9.10(b), the ratio of the switching energy to the charging energy of dot C is between 8 and 16 %.

The following step consists in using a quantum dot as a detector. In Fig. 9.11 the conductance of dot B is plotted as a function of $V_{G_{10}}$ when dots C and D are well defined and isolated from the rest of the 2DEG (panel (a)) or when dots C and D are not defined (panel (c)). We can easily see that plot (a) has a superimposed fine structure with respect to plot (c),

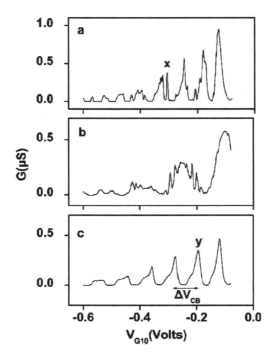

Fig. 9.11.   Detector conductance obtained when dots C and D are defined and sealed
and dot B is used as detector (a), or the series of dots A and B is used as a detector (b). In
(c) the conductance of dot B is plotted when neither dots C or D are defined. Reprinted
with permission from S. Gardelis, C. G. Smith, J. Cooper, D. A. Ritchie, E. H. Linfield,
Y. Jin, *Phys. Rev. B* **67**, 033302 (2003). Copyright 2003, American Physical Society.

which we can readily associate to the switching of electrons between dots
C and D.

In particular, we can focus on the feature denoted with x in Fig. 9.11(c).
In that case, one electron switching between dots C and D shifts the conductance peak by about 20 % of the voltage between two conductance
peaks. In other words, we can say that an electron switching between dots
C and D modifies the charging energy of dot B by 20% of its value.

Again, if all dots are approximately of the same size, we can assume
that such a 20% percentage is always larger than the energy required for
changing the polarization of the dot pair A, B (from previous measurements
on the couple C, D we learned that an energy between 8 and 16% of the
charging energy was sufficient to cause switching of an electron).

In such a way, Gardelis *et al.*[2] have shown that proper operation of AlGaAs-GaAs QCA cells can be achieved.

9.5. Conclusion

The capability of fabricating non-invasive single charge detectors is required for the demonstration of QCA operation. Detectors consisting of a single quantum point contact, biased in order to exhibit a conductance of hundreds of kΩ, provide a strong signal and a relatively easy interpretation.

We have presented simulations of the dot-detector system, which allow us to evaluate the strength of the signals and to optimize the layout for sensitivity. In addition, we have illustrated recent experiments showing that a clever interpretation of detector data, and the use of dots as non-invasive detectors, allows us to extract relevant information on the behavior of systems of multiple dots.

References

1. M. Field, C. G. Smith, M. Pepper, D. A. Ritchie, J. E. F. Frost, G. A. C. Jones, and D. G. Hasko, *Phys. Rev. Lett.* **70**, 1311 (1993).
2. S. Gardelis, C. G. Smith, J. Cooper, D. A. Ritchie, E. H. Linfield, Y. Jin, *Phys. Rev. B* **67**, 033302, (2003).
3. A. W. Rushforth, C. G. Smith, M. D. Godfrey, H. E. Beere, D. A. Ritchie, M. Pepper, *Phys. Rev. B* **69**, 113309 (2004).
4. J. M. Elzerman, R. Hanson, J. S. Greidanus, L. H. Willems van Beveren, S. De Franceschi, L. M. K. Vandersypen, S. Tarucha, L. P. Kouwenhoven, *Phys. Rev. B* **67**, 161308 (2004).
5. G. Iannaccone, C. Ungarelli, M. Macucci, E. Amirante, M. Governale, *Thin solid films* **336**, 145 (1998).
6. S. Selberherr, *Analysis and Simulation of Semiconductor Devices*, Springer Verlag, Wien (1984), p. 23.
7. M. Macucci, A. Galick, and U. Ravaioli, *Phys. Rev. B* **52**, 5210 (1995).
8. M. G. Pala, G. Iannaccone, S. Kaiser, A. Schliemann, L. Worschech, and A. Forchel, *Nanotechnology* **13**, 1 (2002).
9. B. Tanatar and D. M. Ceperley *Phys. Rev. B* **39**, 5005 (1989).
10. M. Macucci, Karl Hess, G. J. Jafrate, *Phys. Rev. B* **48**, 17354 (1993).
11. J. H. Davies, I. A. Larkin, E. V. Sukhourov, *J. Appl. Phys.* **77**, 4504 (1995).

CHAPTER 10

Metal Dot QCA

Gregory L. Snider, Alexei O. Orlov and Ravi K. Kummamuru

Department of Electrical Engineering
University of Notre Dame
Notre Dame, IN 46556, USA

10.1. Introduction

The experimental work presented here is based on a QCA cell using aluminum islands and aluminum-oxide tunnel junctions, fabricated on an oxidized silicon wafer. The fabrication uses standard electron beam lithography and dual shadow evaporations to form the islands and tunnel junctions.[1] A completed device is shown in the SEM micrograph of Fig. 10.1. The area of the tunnel junctions is an important quantity since this dominates island capacitance, determining the charging energy of the island, and hence the operating temperature of the device. For our typical devices the area is approximately 60 by 60 nm, giving a junction capacitance of 400 aF. These metal islands stretch the definition of a quantum dot, but we will refer to them as such because the electron population of the island is quantized and can be changed only by quantum mechanical tunneling of electrons.

10.2. QCA Cell

The first step in the development of QCA systems is a functional QCA cell in which polarization can be switched as a result of the action of a neighboring cell. This confirms the basic premise of the QCA paradigm: that the switching of a single electron between coupled quantum dots can control the position of a single electron in another set of dots.[2] A simplified schematic diagram of our latest QCA system is shown in Fig. 10.2. For

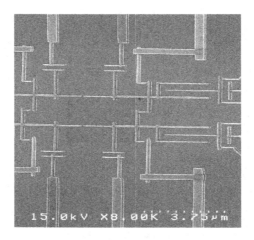

Fig. 10.1.   SEM micrograph of QCA cell and associated electrometers.

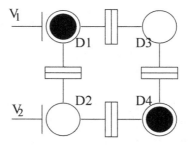

Fig. 10.2.   Schematic of single QCA cell. Electrometers coupled to D3 and D4 are not shown.

clarity, the single-electron transistors (SETs) coupled to D3 and D4 are not shown. The four-dot QCA cell is formed by dots D1-D4, which are coupled in a ring by tunnel junctions. A tunnel junction connects each dot in the cell with an electrode acting as a source or drain. The device is mounted on the cold finger of a dilution refrigerator that has a base temperature of 10 mK, and characterized by measuring the conductance through various branches of the circuit using standard ac lock-in techniques. A magnetic field of 1 T was applied to suppress the superconductivity of the aluminum lines. Full details of the experimental measurements are provided elsewhere.[3]

QCA operation is demonstrated by biasing the cell, tuning the gate voltages, so that an excess electron is on the point of switching between dots D1 and D2, and a second electron is on the point of switching between D3 and

D4. A differential voltage is then applied to the input gates V_1 and V_2 (V_2 = $-V_1$), while all other gate voltages are kept constant. As the differential input voltage is swept from negative to positive, the electron starts on D1, then moves from D1 to D2. This forces the other electron to move from D4 to D3. The experimental measurements confirm this behavior. Using the electrometer signals, we can calculate the differential potential in the output half-cell, $V_{D3} - V_{D4}$, as a function of the input differential voltage. This is plotted in the top panel of Fig. 10.3, along with the theoretically calculated potential at a temperature of 70 mK. Although at a temperature of 0 K the potential changes are abrupt, the observed potential exhibits the effects of thermal smearing, and theory at 70 mK shows good agreement with experiment. In the middle and bottom panels of Fig. 10.3 we report the theoretical excess charge on each of the dots in the input and output half-cells, at 70 mK. This shows an 80% polarization switch of the QCA cell, and confirms the the achievement of the polarization change required for QCA operation.

10.3. Clocked QCA Devices Fabricated Using Metal Tunnel Junctions

As mentioned earlier, clocking in any digital system offers many advantages, and the same happens for the QCA paradigm. Clocking allows us to greatly reduce the power dissipation, control the flow of information, and implement pipelining. In a QCA device clocking is accomplished by modulating the barriers between the dots. In a semiconductor dot system this is easily accomplished by using gates to directly change the barrier between dots.

For QCA arrays implemented in metal islands coupled by capacitors and oxide tunnel junctions, control over barrier heights is much different from that in semiconductor QCAs because oxide barriers cannot be altered by gate potentials. The clocking scheme for the classical Coulomb blockade system with tunnel barriers (single-electron parametron) was suggested by Likharev et al. in Ref. 4 and later refined in Ref. 5. A similar design was used by Toth et al.[6] for clocked control of metallic QCA cells, which we now discuss in more detail.

10.4. Charging Process in QCA Half-Cell

The variable barrier in this case is formed by adding two extra dots to each cell, as shown in Fig. 10.4. To better understand the operation of a clocked

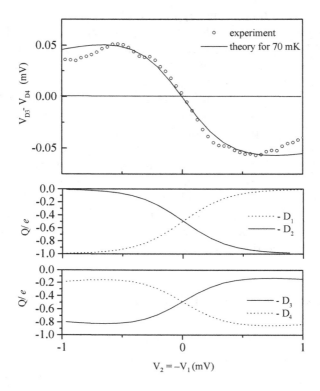

Fig. 10.3. Data from cell measurement. The top panel shows the differential potential of the right half of the cell, while the bottom two panels show the calculated charge on the dots.

QCA cell, we will first consider the operation of a half-cell, or of a QCA latch, delineated by the dashed line in Fig. 10.4. In a half-cell, a modulated barrier is replaced with an extra dot, D2, situated between the two end dots, D1 and D3. Each half-cell is pre-charged with one electron through a TJ (tunnel junction) connected to ground. Calculations of Toth et al.[6] were performed for QCA cells pre-charged with one extra electron. Although our device is initially electrically neutral, any change of the initial charge of the cell is, in fact, equivalent to a fixed offset of the biases V_{IN} and V_{CLK}. Therefore, the considerations of Ref. 6 can also be applied to the initially neutral half-cell).

Each dot of the device is capacitively coupled to a corresponding gate, and the differential input signal (which comes from either external leads,

V_{IN}, or an adjacent latch, e.g. L_1 driving L_2 in Fig. 10.4) is applied to the top and bottom dots. The clocking signal on the middle gate controls the potential on D2, varying the effective Coulomb barrier between D1 and D3. This enables clock control over tunneling within a half-cell. The operational modes of a half-cell are defined by the combination of input and clock biases.

In the null mode, the input signal is zero and the clock bias is such that an electron is localized on D2 (clock LOW). For this combination of gate voltages, the charge state (0,1,0) has the minimal energy configuration. (The numbers in the parentheses are the numbers of excess electrons on D1, D2 and D3 respectively). In active mode, a small input signal is first applied to the top and bottom gates. The magnitude of the input signal, V_{IN}, is small, so it cannot lift the Coulomb blockade for tunneling from D2 to the end dots. The application of the HIGH negative clock signal (with a magnitude $V_{CLK} > e/C_{CLK}$) makes it energetically unfavorable for an excess electron to remain on D2. If no input bias is applied, this electron has equal probability to switch to either of the end dots. However, if a differential voltage V_{IN} is applied to the inputs, Coulomb blockade for tunneling will be lifted only for the electron transfer to the end dot, to which a positive V_{in} is applied. As a result of switching, the charge distribution between the dots reaches its minimal energy configuration. In the locked mode, the electron remains trapped on the end dot to which it was transferred in active mode, regardless of the input signal due to the presence of the clock bias. The excess electron in the locked mode is

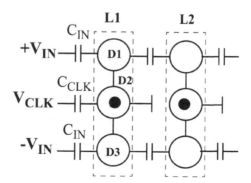

Fig. 10.4. Schematic diagram of a clocked QCA with fixed barriers.[6] The triple dot in the dashed-line box forms a QCA latch. Each latch is pre-charged with one extra electron through the grounded tunnel junction. Two latches form a QCA cell. The null mode is shown.

in the metastable state when the reverse input bias is applied, or in the ground state when the input bias stays the same as in the active mode. In either case, the Coulomb barrier created by the clocking voltage prevents an excess electron from switching. This ability of the half-cell to store an excess electron earns it the name QCA latch, in analogy with a digital D-latch circuit. Several physical mechanisms could result in the escape of the trapped electron from the locked state. One possible source triggering this effect is the thermal excitation of an electron over the Coulomb barrier. However, the probability of this process, $p \approx \exp(-E_B/(kT))$, is small if $E_B \gg kT$. Here, $E_B \approx e^2/C_J$ is the height of the Coulomb barrier and C_J is the junction capacitance (we assume C_J to be the dominant term in the total capacitance of the dots). For our device $E_B \approx 0.5$ meV, so that, for $T = 70$ mK, $E_B \gg kT \approx 6$ μeV, and direct tunneling is strongly suppressed. In that case, the second-order processes (cotunneling)[7] are responsible for the escape of the trapped electrons. The average lifetime of an electron in the metastable state τ can be estimated using the zero-temperature approximation[7] for a line of N tunnel junctions biased with a potential difference V:

$$\tau = 2R_j C_j [(N-1)!]^2 N^{-2N} (2N-1)! \left(\frac{\pi^2 R_j}{R_q}\right)^{N-1} \left(\frac{E_C}{eV}\right)^{2N-1}. \quad (1)$$

For $R_J = 200$ kΩ, $C_J = 0.3$ fF, $V = 250$ μV, we arrive at an estimate for a retention time, for a latch with two such junctions, $\tau \simeq 100$ ns, and in the latch with six TJ connected in series, τ exceeds 10^6 s.

Following Ref. [6] we will first consider the equilibrium charging of the QCA half-cell,[8] describe the physical mechanisms of single-electron transfer in a half-cell and then proceed with the analysis of its operation in a QCA circuit. Equilibrium charging takes place in a half-cell when the trapping time of an excess electron in the metastable state of the locked mode is much shorter than the clocking period. In our experimental setup this is the case for a half-cell with two TJs, because even the shortest time constant used for data acquisition, $\tau_{acq} \simeq 10^{-4}$ s $\gg \tau \simeq 10^{-7}$ s. Therefore no latching can be observed, and the electron distribution on the dots reaches an equilibrium ground state for every measured data point.

The half-cell with two Al/AlO$_x$ junctions separating three Al dots D1-D3 is shown in Fig. 10.5. Each dot is also coupled to the SET electrometers E1, E2 and E3. Three leads capacitively coupled to the dots act as signal $(+V_{IN}, -V_{IN})$ and clock V_{CLK} inputs. To study the operation of the QCA half-cell, we analyze the single-electron switching processes (measured by

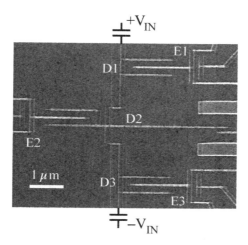

Fig. 10.5.   SEM micrograph of a QCA half-cell with two TJs separating three dots with three SET electrometers. Input gates are shown schematically.

SET electrometers) as a function of the input and clock biases and compare with theory.[5,6] For characterization of the half-cell we measure charging phase diagrams (CPD) — gray-scale maps of the electrostatic potential on the dots of the latch plotted versus input and clock biases. White color in the plot corresponds to positive potential and dark color corresponds to negative potential on the dot to which the electrometer was coupled. The maximum potential swing (from black to white) in the experiments described below was about 200 μV. Figure 10.6 shows an equilibrium CPD measured by electrometer E3. Superimposed with the experimental data are the results of theoretical calculations[6] showing the borders between ground state charge configurations.

For the calculations, we use experimentally extracted capacitances with offset background charge as the only adjustable parameter. Due to the random background charge, the whole picture is offset with respect to zero, thus the point of neutrality, N, is not positioned exactly at zero gate voltage. The numbers of excess electrons shown in the brackets correspond to the local (within the area delineated by hexagons) ground states. Negative numbers mean that electrons are missing from the respective dots. The center of each hexagon corresponds to a monostable (null) state.

For the ideal case (no random background charge) with no voltages applied to the gates, the charge configuration (0,0,0) has the lowest electro-

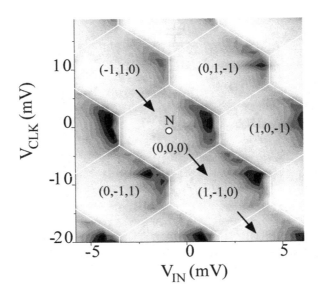

Fig. 10.6.   CPD of the half-cell made with two TJ separating the dots, measured by an electrometer attached to D3 in Fig. 10.5. The differential bias, V_{IN}, is applied as shown in Fig. 10.4. Numbers in brackets are the excess electrons on the top, middle, and bottom dots, respectively.

static energy because of charge neutrality. For classical metal systems with continuous energy spectra, the CPD is e-periodic, and similar energy considerations are applicable to each hexagon in Fig. 10.6 (the offset voltages compensate excess charges on the dots). Note that there are no transitions (resulting in change in brightness) along the directions of the arrows. The charge on D3 remains constant along this direction. The absence of a monotonic slope along the scanned voltage axes in the phase diagrams results from a charge cancellation technique used in the experiment. It cancels out the effect of a monotonic change of the electrostatic potential on the half-cell and thus allows the observation of events of interest only — the single-electron transitions.

Now that we have reviewed the general properties of charging in a QCA half-cell, let us now concentrate on a specific region of the CPD to demonstrate clocked control. Figure 10.7 shows a small region of the CPD measured simultaneously by three electrometers E1-E3 in all three dots of a half-cell. Using the CPD, we will trace single-electron transitions within the half-cell. Here and below we consider the clock level HIGH to be negative, though a positive clock signal could equally be used. In this case, an excess positive charge (hole) on the end dots is used to encode the binary information.

We start at the neutrality point, $V_{CLK} = 0$, $V_{IN} = 0$ (the CPD is offset to place it at $V_{CLK} = V_{IN} = 0$ for clarity). For $V_{CLK} = 0$, changing the input signal from $V_{IN} = -0.2$ mV to $V_{IN} = 0.2$ mV (path from a to c in Fig. 10.7) does not cause a single electron transfer, and the system remains in the same charge state $(0,0,0)$. In the vicinity of this point the charge configuration $(0,0,0)$ has the lowest energy. This is the null mode, where a cell remains unpolarized even when a small input signal is applied. A change in the clocking bias toward clock HIGH polarizes the half-cell. For instance, if the input bias is set to $V_{IN} = -0.2$ mV and the clock bias changes from $V_{CLK} = 0$ mV to $V_{CLK} = -6$ mV, then D3 gains an electron, and this electron comes from D2, leaving a hole (missing electron) behind (path from a to c in Fig. 10.7). Thus a transition $(0,0,0)$ to $(0,-1,1)$ occurs along this trajectory. Note that the number of electrons on D1 does not change, though the potential on that dot becomes more positive. Similarly, by moving from b to d, a transition from the $(0,0,0)$ to the $(1,-1,0)$ state is accomplished. The transition from the null into the active mode occurs at the dotted line when the clocking signal changes the charge configuration in the half-cell, where the final state is determined by the polarity of V_{IN}.

Further change of the clock signal toward c (or toward d) brings a half-cell into the locked mode. Here it is energetically unfavorable for an electron to move to D2, so D2 keeps its positive net charge, and an electron stays either on D1 or D3, depending on the polarity of the input bias in the active mode. If the polarity of the input signal is reversed, the electron is trapped in a metastable state,[5,6] which has higher energy than a true ground state. As discussed above, the trapping time for a half-cell with two TJs, shown in Fig. 10.7, is much smaller than the acquisition time and the metastable state in the locked mode cannot be observed for the conditions of this experiment.

To summarize, the analysis of the CPDs in Figs. 10.6 and 10.7 confirms the results of theoretical calculations,[5,6] and shows that single electron switching can be accomplished by clock signals with switching direction defined by the input. The CPD in Figs. 10.6 and 10.7 correspond to the equilibrium ground state, and do not show any metastability within the experimental time resolution. To demonstrate latching, the lifetime of an electron in the metastable state should be made significantly longer. This can be achieved, for example, using multiple TJs to connect the dots. In our experimental setup the the upper limit of the bandwidth for measurements of the SET electrometers is about 10 kHz, which sets the lower limit for a clock cycle at which we can achieve an adequate temporal resolution $T_{CLK} \approx 10^{-1}$ s. For six junctions this creates favorable conditions for QCA latch operation at low clocking frequencies, such that $T_{CLK} \ll 10^6$ s, which was estimated for zero temperature using Eq. (1).

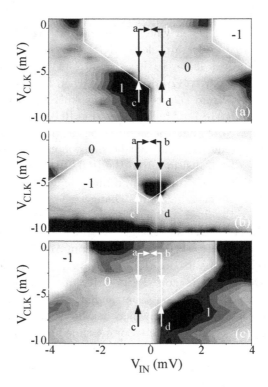

Fig. 10.7. Equilibrium CPD of the clocked QCA half-cell, measured by electrometers E1-E3. (a) D3; (b) D2; (c) D1. Numbers on the graphs represent the number of excess electrons on the respective dots within the areas confined by the dashed lines. Dashed lines on the plots are calculated using the modeling algorithm described in Ref. [6] and define the borders between equilibrium ground state charge configurations. Adapted with permission from A. O. Orlov, I. Amlani, R. K. Kummamuru, R. Ramasubramaniam, G. Toth, C.S. Lent, G. H. Bernstein, and G. L. Snider, *Appl. Phys. Lett.* **77**, 295 (2000). Copyright 2000, American Institute of Physics.

10.5. QCA Latch Operation

The SEM micrograph of a QCA latch with multiple tunnel junctions (MTJs) is shown in Fig. 10.8. It consists of three Al dots D1, D2 and D3 connected in series by MTJs. Each MTJ in turn consists of three tunnel junctions and two small extra islands. To minimize parasitic coupling to these islands from the gates, the area of the islands in MTJs ($0.3 \times 0.08\ \mu m^2$) is made much smaller than the area of dots D1-D3 ($5 \times 0.08\ \mu m^2$). To detect the electron transfers in the half- cell, the end dots are coupled to SET electrometers E1 and E2.

Figure 10.9 shows a CPD for two scan directions of the input bias. The areas within the hatched triangles correspond to metastable states of electrons in the half-cell. In the upper panel of Fig. 10.9 the region of metastability expands to the right from the equilibrium border, and in the lower panel it expands to the left. This occurs because the Coulomb barrier caused by the HIGH clock level suppresses the switch to the new minimal energy configuration even if the equilibrium border is crossed.[6] Note that the point of neutrality is offset from zero by 1.8 mV along the input voltage axis and by −0.5 mV along the clock bias, due to the random background charge. Here and below in the text, the differential input bias, and clock signal are considered to be applied relative to this initial offset point. Figure 10.10(a) illustrates the experiment in the upper panel of Fig. 10.9. Here, we plot the energy diagrams as the input differential voltage in Fig. 10.9 changes from negative to positive. Corresponding points, α, β, γ are shown in the upper part of Fig. 10.9. At point α, where the input voltage $\delta = -0.3$ mV

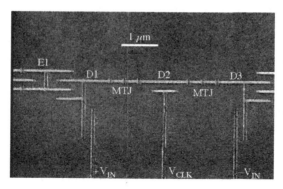

Fig. 10.8. SEM micrograph of a QCA latch fabricated with multiple tunnel junctions (MTJ) separating the dots. Note three junctions and two extra islands in each MTJ.

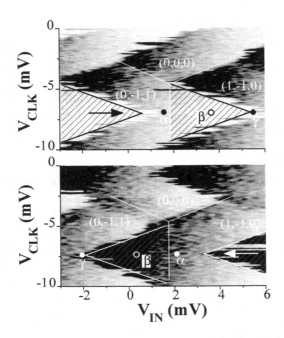

Fig. 10.9.   QCA latch charging phase diagrams measured by E_1. To obtain the plot, the input differential bias is scanned in the direction shown by the arrows. The clock signal is stepped with increments of 100 μV. Numbers in brackets are the ground state charge configurations on D1, D2, and D3. Hatched triangles mark the areas of metastability. Equilibrium borders separating the charge configurations in the area of interest are shown as thick lines.

(the equilibrium border is $\delta = 0$), an excess electron on D3 is in the ground state, because (0,-1,1) is the minimal energy configuration. As the input bias increases to point β ($\delta = +1.5$ mV), the minimal energy configuration changes to (1,-1,0), but the Coulomb barrier prevents the excess electron from switching, so it remains trapped on D3. Finally, at $\delta = +3.4$ mV (point γ), the two electron configurations (0,-1,1) and (0,0,0) become energetically equivalent. As a result, there is no Coulomb barrier any more to hold an excess electron, and, with a slight increase in δ, it switches from D3 to D1. At this point, charge configuration (0,-1,1) has changed to (1,-1,0).

The lower panel of Fig. 10.9 illustrates the charging process for the input bias scan in the opposite direction. In this case for the point at $\delta = +0.3$ mV (α), the configuration (1,-1,0) with an electron on D1 has the lowest energy, metastability exists at $\delta = -1.5$ mV (β), and an electron is pushed from D1 to D3 at $\delta = -3.4$ mV (γ). It is clear that only when the combination

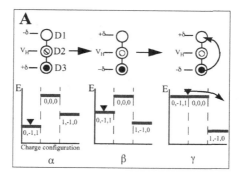

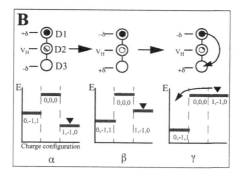

Fig. 10.10. Switching process in a triple dot for two directions of the input bias scan shown in Fig. 10.12 with clock voltage set HIGH. E is the electrostatic energy of the configuration, δ is the input differential signal. A black circle represents an excess electron, and a hatched circle represents a missing electron. Black triangles in the energy diagrams mark the current charge configuration of the system.

of voltages on the gates is such that the Coulomb barrier separating the two charge configurations is eliminated, an electron can switch into a true ground state. The electrostatic potentials on D1 and D3 in these regions, therefore, show hysteresis for the forward and reverse scans of the input bias (Fig. 10.11). The lower (upper) branch of the hysteresis loop corresponds to an electron trapped on D1 (D3).

Let us look now at the charging events occurring for changing clock and fixed input voltages, as these are the conditions for QCA latch operation. For this purpose we measure the CPD with the clock signal scanned for different settings of the input differential bias (Fig. 10.12). Note that this CPD differs dramatically from that in Fig. 10.9, and resembles that of

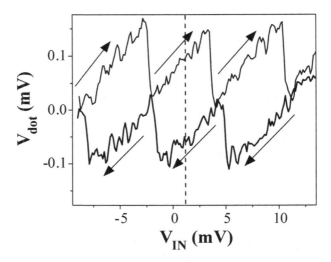

Fig. 10.11.   Signal in the detector $E1$ for two directions of the scan in the bistable region
($V_C = -7.5$ mV). The dashed vertical line delineates the position of an equilibrium
border between charge configurations in Fig. 10.12. Arrows indicate the scan direction.

Fig. 10.7. As before, the initial (LOW) clock bias is set in the monostable
region of the map (point N in Fig. 10.12). With an input at points E or
B, as the clock bias changes, a transition from monostability to bistability
takes place, and at the same time a bit represented by a switching electron
gets stored in the latch.

Figure 10.13(a) illustrates single-electron transitions occurring along the
path B-C-D in Fig. 10.12. Here, at point B, $V_{IN+} = -V_{IN} = +\delta = 0.5$ mV
is applied while the clock bias is low, $V_{CLK} = V_L = -1$ mV (in the ideal
case with no background charge, this value is 0 mV). The ground state
configuration remains (0,0,0) because the input bias is small and the latch
is still in the monostable null state. As the clock voltage changes, at a
transitional point C ($V_{CLK} = V_T = -4$ mV) the two charge configurations
(0,0,0) and (1,-1,0) have the same energy, thus the transition (0,0,0) (1,-
1,0) is allowed. From C to D the charge configuration (1,-1,0) is more
energetically favorable.

Further change of the clock voltage toward $V_{CLK} = V_H = -7.5$ mV
leads to energy separation of charge configurations (0,0,0) and (1,-1,0). At
point D the minimal energy configuration is (1,-1,0), and D1 and D3 are
now separated by a Coulomb barrier. The electron is now locked on D1
even if the input bias, δ, is set to zero, or even reversed, because, in order

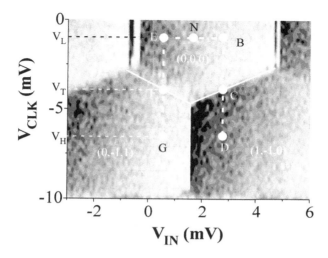

Fig. 10.12. `Latch CPD for the clock scans measured by E1. Input differential bias is stepped in 100 μV increments. The scan direction is shown by the arrow. Equilibrium borders between states in the area of interest are shown as dark dashed lines.

for an electron to switch to D3, the system should go through the charge state (0,0,0) which now has much higher energy. This case is illustrated in Fig. 10.12(b). Note that the latch is most stable when point D is situated along the vertical midpoint of the triangle in Fig. 10.9(b), since that represents the condition of largest Coulomb barrier. Further change of the clock voltage will reduce the barrier due to periodicity of the electrostatic energy of the system.[5,6,9]

We may easily visualize the behavior of the CPD for D3 relative to that of D1 as shown in Fig. 10.12. To do this, merely flip Fig. 10.12 about the middle vertical axis, i.e. the equilibrium border. The potential on D3 behaves like the path EFG while the potential on D1 follows the path BCD. No abrupt transition on D3 occurs, but rather a monotonic increase in potential is observed. This is because unlike in D1(D2) which gains (loses) an electron and its potential decreases (increases) abruptly at point C, further increases in V_{CLK} through point F do not cause a switch in D3, even though its potential continues to increase. Thus, the difference in potentials between D1 and D3 provides the inputs for driving subsequent stages in a line of QCA latches. Figure 10.13(b) illustrates the switching process for inverted input voltage ($\delta = -0.5$ mV): the respective points E-F-G show how the charge configuration changes from (0,0,0) to (0,-1,1).

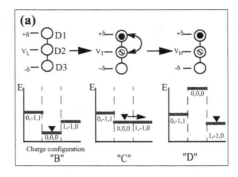

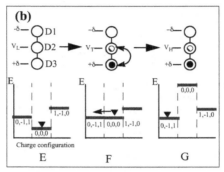

Fig. 10.13.   Single-electron transitions in QCA latch operation. (For legends see Fig. 10.10). Note that δ is not changing throughout this sequence.

Analysis of Figs. 10.9 and 10.12 shows that in order for a QCA latch to function, the input signal should be confined within the area of the two hatched triangles containing β-points in Fig. 8, while the clock signal should change by the amount corresponding to the vertical size of the hexagons in Fig. 10.6 to maximize the Coulomb barrier separating D1 and D3. Thus, by analyzing the CPD in Figs. 10.9 and 10.12, one can find appropriate magnitudes of the clock and input signals and choose the working point for latch operation.

Let us now consider the operation of the QCA latch in the time domain for two polarities of the input signal (Fig. 10.14). At $t_1(t_5)$ the input signal is applied. In Fig. 10.12 this corresponds to the deflection from the null point N to the point $E(B)$, resulting only in a small variation of the potential in D1, proportional to V_{IN}. As the clock voltage is set HIGH (note that CLK LOW=0, and CLK HIGH=-6 mV) at $t_2(t_6)$, it pushes an electron from

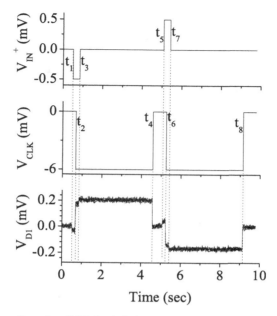

Fig. 10.14. Operation of a QCA latch.[9] A small input signal defines the direction of switching in the latch, while the clock signal triggers the switching. The latch retains the bit as long as the clock is applied. Five successive traces are shown to outline the noise levels.

D2 to D3(D1). This results in a large shift of the electrostatic potential on D1(D3). The resulting potential difference between D1 and D3 provides a signal to drive an adjacent latch which is activated by a separate clock line. Once the clock is set HIGH, the input signal can be removed, and the electron remains trapped on D3(D1), until the clock is returned to low at t_4 (t_8). Thus, a cycle describing the operation of a latch based on a QCA half-cell is as follows: null $\rightarrow$ active $\rightarrow$ locked $\rightarrow$ active $\rightarrow$ null.

Figure 10.15 demonstrates the ability of the latch to retain the stored electron in the locked phase, regardless of the input signal (error rejection).[10] Here input voltages of both polarities with magnitude four times greater than the initial input bias are applied to the input gates after the clock is set high. The electron remains latched, regardless of the input signal (note that the input signal should stay within the area of hatched triangles in Fig. 10.9. This defines the maximum disruptive input signal to be rejected by the latch). A small change in the amplitude of the output voltage, which is seen in Fig. 10.15, is the result of the sawtooth shape of the dot potential (Fig. 10.11).

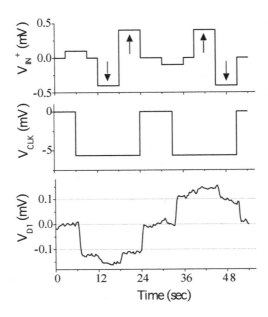

Fig. 10.15. Demonstration of a QCA latch error rejection ability for the two input polarities. Once the latch is set into a locked state, the disruptive input signal, marked by arrows, does not affect the state of the latch.

One of the most important parameters which determine the success of any logic device is the speed of switching for binary operation. The operational speed of the latch is determined primarily by the tunneling time of an electron ($\tau \approx R_J C_J \approx 10^{-10}$ s, where $R_J \approx 10^6$ Ω and $C_J \approx 10^{-16}$ F are the resistance and the capacitance of the junction, respectively). For quasiadiabatic operation the switching speed needs to be reduced by an order of magnitude.[11] Thus, for the current Al/AlO$_x$ QCA prototype, the estimate for the switching speed limit is on the order of 1 ns. For future molecular implementations due to much smaller total capacitance ($C \approx 10^{-19}$ F) the expected switching speed is on the order of picoseconds.

It is worth noting that the maximal clock speed in our current experiment is limited not by the switching speed in the latch, but by parasitic RCs in the electrometer circuits. Since the temporal resolution of the SET readout is limited as a result to about 0.1 ms, any events occurring at a higher rate simply are not detected. To solve this problem a radio frequency SET[12] electrometer will be used in the future experiments.

10.6. Two Stage QCA Shift Register — a Clocked QCA Cell

Above we described the operation of a building block for clocked QCA — a latch. Next, the operation of a coupled system where one latch drives the other by means of single electron switching, powered by the clock is considered: a shift register (SR), comprising a clocked QCA cell, is considered. The simplified circuit diagram and an SEM micrograph of the device are shown in Fig. 10.16. The device consists of two QCA latches and two readout electrometers E1 and E2. The two latches are capacitively coupled to each other using lateral capacitors CC. Each QCA latch consists of three dots D1-D3, and D4-D6, separated by MTJs. SET electrometers E1 and E2 are used to measure the state of each latch. Operation of the QCA shift register is performed using a two-phase clock (CLK1 and CLK2). The details of the operation of the two-stage SR in the time domain are shown in Fig. 10.20.[13] First, the differential signal V_{IN} corresponding to logical 0 (logical 1) is applied to the inputs V_{IN}^+, and V_{IN}^- at t_1 (t_7) (Fig. 10.17(a)). As described above, L1 remains in the monostable null state until CLK1 is set HIGH at t_2 (t_8) (Fig. 10.17(b)). When CLK1 is activated, it causes a transition of an electron in L1 (Fig. 10.17(c)). After that, the signal input is removed at t_3 (t_9) and the state of L1 no longer depends on the input signal. Then CLK2 is applied to L2 at t_4 (t_{10}) (Fig. 10.17(d)), and

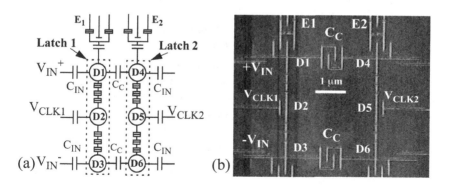

Fig. 10.16. Circuit diagram and a micrograph of a two-stage QCA Shift Register. Adapted with permission from A. O. Orlov, R. Kummamuru, C. S. Lent, G. H. Bernstein, and G. L. Snider, *Surface Science* **532-535**, 1193 (2003). Copyright 2003, Elsevier Science B. V.

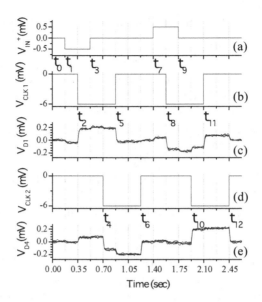

Fig. 10.17. Operation of a QCA shift register. Two phase-shifted clock signals are applied to two capacitively coupled latches to shift binary information from one latch to the next in a sequential manner controlled by the clock. Five successive traces are shown. Adapted with permission from A. O. Orlov, R. Kummamuru, C. S. Lent, G. H. Bernstein, and G. L. Snider, *Surface Science* **532-535**, 1193 (2003). Copyright 2003, Elsevier Science B. V.

an electron in L2 switches in the direction determined by the state of L1 (Fig. 10.17(e)). L2 holds the bit after CLK1 is removed at t_5 (t_{11}), as long as CLK2 is high (until t_6 (t_{12})). The cycle describing the operation of a QCA SR is as follows: monostable → input applied → CLK1 applied and L1 is active → input removed → CLK2 applied and L2 is active → CLK1 is removed. At this time, L1 becomes inactive and is ready to receive new information. The information encoded in the position of a single electron is shifted to L2 and stored there. It is thus possible to see that a clocked QCA cell (or QCA SR) operates as predicted in Ref. 5, 6.

10.7. Simulation of a Multi-Stage Shift Register

The operation of any clocked QCA circuit would require synchronous operation of multiple cells to process the binary information. Hence, the performance of a multi-stage SR, especially with regard to preservation of logic levels and vulnerability to digital errors, is of considerable interest. The current device can be used to replicate the propagation of a single bit through such an SR. In a multi-stage SR, a bit is first written into the circuit by the input and then moved along the circuit using each latch as an input to the next (Fig. 10.18(a)). The same situation can be simulated using the two-stage SR by moving the bit back and forth from one latch to the other (Fig. 10.18(b)). Initially, a bit is written into the first latch by the input. Then, using L1 as input, the bit is copied into L2, after which L1 is turned off. Then using L2 as input, the bit is copied back into L1, and L2 is turned off. This process can be repeated a number of times to achieve the same effect as transferring a bit through a long line of latches.

Figure 10.19 shows the timing diagram of the experiment performed for 5 cycles.[13] Initially, all the signals are zero and the two latches are in the neutral state. Once the input (binary 1) and clock signals are applied to L1, it switches. The input is then removed and the bit is stored in L1. The clock signal is then applied to L2, and it switches using L1 as its input. L1 is then switched off, and the bit is now stored in L2. Instead of applying the clock signal to a third latch in the line, it is applied to L1 which sees L2 as an input and switches accordingly. Then L2 is switched off and the bit is stored once again in L1. This cycle is repeated 5 times to simulate an SR made of 11 latches. In the second half of the experiment, this scheme is repeated with an input of the opposite sign (binary 0). This is necessary to verify that bit inversion at the input indeed leads to the bit inversion at the output. The above experiment demonstrates that the direction of the information flow in the circuit is controlled by the sequence of clock signals applied to latches. Another observation that can be made from Fig.10.19 is that, although the input is applied only once at the beginning of the cycle, we do not see any degradation in the voltage levels as the bit is moved back and forth between the latches, indicating that signal levels would be preserved in multi-stage QCA SRs. The major reason for this stems from the ability of clocked QCA to exhibit power gain (see below) — in a QCA circuit, the clock signal provides the energy needed for digital level restoration, just as a conventional power supply does it for FET (Field Effect Transistor) circuits.

10.8. QCA Power Gain

Power gain is an important requirement for any real electronic device. Power gain makes possible logic devices which restore signal levels and overcome noise in a system. Without power gain, the signal energy put into the system by the inputs is quickly lost to the environment. Power gain in digital

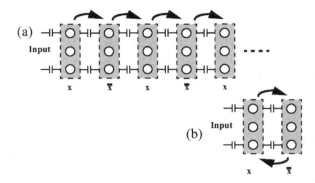

Fig. 10.18. Using a two-stage SR to simulate a multi-stage SR. (a) In a multi-stage SR, a bit is moved sequentially in a single direction from one latch to the next. The bit is inverted at each step. (b) The two-stage SR can be used to simulate a longer SR by moving the bit back and forth from one latch to the other instead of moving it in a single direction.

logic devices is not the same as in linear amplifiers, because logic devices are saturating amplifiers. In a system of saturating amplifiers, net power gain occurs only when a weak signal is applied to the input. If a strong signal is applied, the output is equal to the input and the power gain is unity. In conventional digital logic, if a weak signal is applied to a gate, power is drawn from the voltage supply to produce an output signal with the full logic voltage. In a QCA cell, the clock line plays the role of the voltage supply in providing the power to restore the signal. In a QCA system, a weak signal could be caused by loss of energy in the system, an abnormal capacitor, or by a latch with a low charging energy. However, a small input signal is sufficient to decide the direction of switching while the clock provides most of the energy for the switching. Thus, when a weak input occurs, the clock provides the energy required to switch the latch and restore the logic level.

In our demonstration of QCA power gain, we must first make clear our definition of power gain, which is the ratio of the power delivered by a cell to the power applied to that cell, as shown in Eq. (2), which also relates the power gain to the work done by and on the cell divided by one clock period. Using this definition, we will demonstrate power gain by measuring the work

$$\text{Power Gain} = \frac{P_{\text{out}}}{P_{\text{in}}} = \frac{W_{\text{out}}/T}{W_{\text{in}}/T} \tag{2}$$

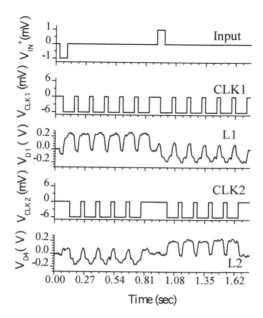

Fig. 10.19. Experiment to simulate a multiple stage SR.[13] The input is applied only at the beginning, to write a bit into the SR. Then the bit is transferred from one latch to the other for five cycles. In the second half of the experiment, the sequence of events is repeated with an input of the reversed polarity. This experiment can be used to investigate the propagation of errors in a multi-stage SR. Adapted with permission from A. O. Orlov, R. Kummamuru, C. S. Lent, G. H. Bernstein, and G. L. Snider, *Surface Science* **532-535**, 1193 (2003). Copyright 2003, Elsevier Science B. V.

done on a cell by the input over one clock cycle as well as the work done by the cell on the next cell, and compare the two. The work done is defined by Eq. (3), where V is the voltage at a cell lead, and Q is the charge at that connection.

$$W = \int V dQ. \tag{3}$$

In the experiment, the work is measured by measuring the lead voltage, and since the leads to the cell connect through capacitors, the charge is calculated through the voltage on each side of the capacitor and the value of the capacitor. Over one clock cycle, a plot of V vs. Q forms a loop whose area is the work done, and the direction in which the loop is completed denotes whether work is done by the cell or on the cell. Clockwise denotes work done on the cell, and counterclockwise work done by the cell.

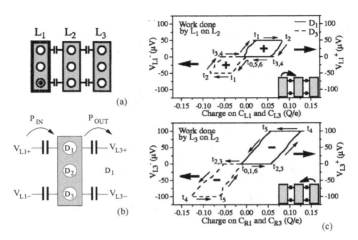

Fig. 10.20. (a) Simplified schematic of QCA shift register. (b) Simplified schematic of power gain experiment. (c) Measurement of work done by and on latch L2, showing a power gain of 2.07. Adapted with permission from R. K. Kummamuru, J. Timler, G. Toth, C. S. Lent, R. Ramasubramaniam, A. O. Orlov, G. H. Bernstein, and G. L. Snider, *Appl. Phys Lett.* **81**, 1332 (2002). Copyright 2002, American Institute of Physics.

In the experiment a single latching cell will be used (L2), and power supplies will be used to simulate a shift register with an input cell to the left (L1), and an output cell to the right (L3), as shown in Fig. 10.20(a). The voltages V_{L1+}, V_{L1-}, V_{L3+} and V_{L3-} (input and output, in short) are applied to capacitors C_{L1}, C_{L3}, C_{R1} and C_{R3}, respectively to simulate latches L1 and L3 as shown in Fig. 10.20(b). To determine the magnitudes of V_{L1} and V_{L3}, the dot potential swing in latch L2 is first measured and then a smaller signal is used to simulate a weak latch L1, while a signal of the same magnitude is used to simulate a normal latch L3. Figure 10.20(c) shows the Q-V plots for the signals V_{L1} and V_{L3} over one clock period simulating bit motion in the shift register. Full details of the experiment are given elsewhere.[14] The $Q-V$ plot for V_{L1} (the top panel of Fig. 10.20(c)) shows a clockwise direction indicating that latch L1 performs work on latch L2. However, the $Q-V$ plot for V_{L3} (the bottom panel of Fig. 10.20(c)) shows a counterclockwise direction indicating that latch L2 performs work on latch L3. Hence, work is being performed in the same direction as the transfer of the bit. The ratio of the area enclosed by the plot in the top panel to the area enclosed by the plot in the bottom panel gives the power gain. This ratio calculated from Fig. 10.20(c) is 2.07. This demonstrates that clocked QCA cells can provide true power gain.

References

1. T. A. Fulton, and G.H. Dolan, *Phys. Rev. Lett.* **59** 109 (1987).
2. A. O. Orlov, I. Amlani, G. H. Bernstein, C. S. Lent, and G. L. Snider, *Science* **277**, 928 (1997).
3. I. Amlani, A. O. Orlov, G. L. Snider, C. S. Lent, and G. H. Bernstein, *Appl. Phys. Lett.* **72** 2179 (1998).
4. K. K. Likharev, and A. N. Korotkov, *Science* **273**, 763 (1996).
5. A. Korotkov, and K. K. Likharev, *J. Appl. Phys.* **84**, 6114 (1998).
6. G. Toth, and C. S. Lent, *J. Appl. Phys.* **85**, 2977 (1999).
7. D. V. Averin, and A. A. Odintsov, *Physics Lett. A* **140**, 251 (1989).
8. A. O. Orlov, I. Amlani, R. K. Kummamuru, R. Ramasubramaniam, G. Toth, C.S. Lent, G. H. Bernstein, and G. L. Snider, *Appl. Phys. Lett.* **77**, 295 (2000).
9. A. O. Orlov, R. Kummamuru, R. Ramasubramaniam, C. S. Lent, G. H. Bernstein, and G. L. Snider, in the *Proceedings of the 1st IEEE Conference on Nanotechnology*, Maui, HI (2001).
10. A. O. Orlov, R. K. Kummamuru, R. Ramasubramaniam, G. Toth, C. S. Lent, G. H. Bernstein, and G. L. Snider, *Appl. Phys. Lett.* **78**, 1625 (2001).
11. C. S. Lent, and P.D. Tougaw, *Proc. of the IEEE* **85**, 541 (1997).
12. R. J. Schoelkopf, R.J., P. Wahlgren, A. A. Kozhevnikov, P. Delsing, and D. E. Prober, *Science* **280**, 1238 (1998).
13. A. O. Orlov, R. Kummamuru, C. S. Lent, G. H. Bernstein, and G. L. Snider, **Surface Science, 532-535**, 1193 (2003).
14. R. K. Kummamuru, J. Timler, G. Toth, C. S. Lent, R. Ramasubramaniam, a. O. Orlov, G. H. Bernstein, and G. L. Snider, *Appl. Phys Lett.* **81**, 1332 (2002).

CHAPTER 11

Molecular QCA

Craig S. Lent

Department of Electrical Engineering
University of Notre Dame
Notre Dame, IN 46556, U.S.A

11.1. Introduction

Creating QCA cells composed of single molecules holds enormous promise for realizing the ultimate limits of electronic device miniaturization and integration.[1] In contrast to present metal-dot QCA cells, the small size of a molecule means that Coulomb energies are much larger, so room temperature operation is possible. In contrast to transistors, the power requirements and heat dissipation of QCA are low enough that high-density molecular logic circuits and memory are feasible. In contrast to all lithographic device fabrication techniques, which introduce variations in device characteristics, each molecular cell can be made exactly identical using chemical synthesis.

Candidate QCA molecules must be designed to exhibit several basic characteristics:

1) Areas of the molecule must localize charge to form "dots." The role of the dot in molecular QCA is played by molecular redox centers with the bridging ligands between redox centers providing the tunneling barrier which causes the localization. The extent of localization in an unbiased molecular cell is not as critical as the localization when the cell is responding to a bias from another molecular cell.

2) Charge must be able to tunnel between dots to enable switching. Upper limits on the switching speed are determined by the tunneling time through the bridging ligands. The effective barrier can be chemically varied to make this transfer time as small as seconds, or as fast as motion within a

single extended orbital (on the order of 10^{-15} sec.). Electron transfer rates are known to extend to the THz regime, though direct experiments on field-driven transfer have not been done and are an area that needs further investigation.

3) The field from one molecule must be sufficient to switch the state of a neighboring molecule. The signature of this is seen in the cell-cell response function, which is realized as a plot of the dipole (or quadrupole) moment of a target molecule as a function of the dipole (or quadrupole) moment of a neighboring driver molecule. A clearly nonlinear behavior with saturation at fully polarized states is the key requirement.

In addition to these basic requirements, there are other issues. We need to be able to controllably anchor the molecules on a surface so that circuit layouts can be formed. This is likely to be best accomplished by covalent bonds. Clocked control is necessary and can be accomplished by having middle "null" dots within the molecule, or by simply moving the charge reversibly from the molecule to the substrate. In any case charge neutrality of the molecule+substrate must hold. Either the molecule is neutral (which would likely be preferable), or the molecule is an ion with a neutralizing charge in the form of an image charge in the substrate.

Thus far the Notre Dame group has investigated theoretically and experimentally three different candidate molecular systems. (1) The Aviram molecule and its related forms, (2) a two-dot molecule with a ferrocene center and a ruthenium center forming the dots, and (3) a four-dot square molecule with ferrocene groups forming the dots.

11.2. Aviram's Molecule: A Simple Model System

A simple two-dot molecule was proposed by Aviram, in the context of field-driven molecular switches, and studied further by Hush.[6] In the Aviram molecule the dots are formed from allyl groups on the ends of alkane chains (Figs. 11.1 and 11.2). The advantage of this system is that it is very simple — only carbon and hydrogen atoms. Though the molecule is stable, the radical end groups are very reactive so it plays the role of a model system rather than a realistic candidate. It further lacks the functional groups that would enable surface-binding, orientation, and ordering. Nevertheless it is very helpful in sorting out the features of real molecules that are important for QCA operation.

Quantum chemistry modeling[7] of the Aviram molecule shows that indeed a molecule can exhibit QCA bistability and molecule-molecule cou-

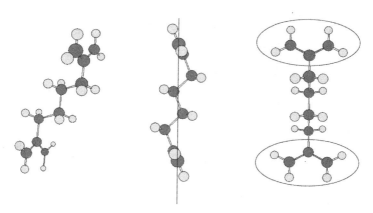

Fig. 11.1.   Three views of the Aviram molecule. The allyl groups acts as dots forming a two-dot molecular QCA cell.

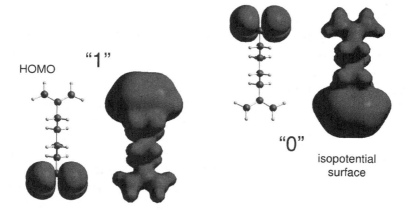

Fig. 11.2.   The highest occupied molecular orbital (HOMO) and isopotential surface for the two stable states of the Aviram molecule.

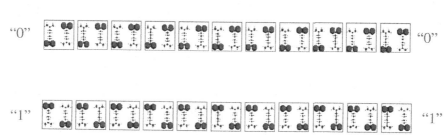

Fig. 11.3.   QCA cells encode information in the charge configuration as shown in (a) and (b). The response of one cell to its neighbor is very nonlinear (c).

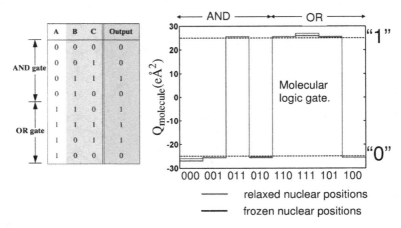

	A	B	C	Output
AND gate	0	0	0	0
	0	0	1	0
	0	1	1	1
	0	1	0	0
OR gate	1	1	0	1
	1	1	1	1
	1	0	1	1
	1	0	0	0

Fig. 11.4. Calculated response of QCA majority gate realized with Aviram molecule pairs.

pling. When driven by a field, the dipole moment of the molecule flips from one sign to the other abruptly — thus displaying the requisite bistability. Moreover the dipole moment of one molecule induces a dipole moment in a neighboring molecule in a very nonlinear way. Arrays of molecules forming QCA binary wires (Fig. 11.3) and majority gates (Fig. 11.4) have also been shown to work within the quantum chemistry calculations.

The Aviram molecule is simple enough that it can be used to compare the usefulness of different quantum chemistry approximations. Local density calculations turn out to drastically underestimate the degree of localization and so prove poor. Unrestricted Hartree-Fock levels yield agreement with far more exact treatments. It is important to realize that, to capture the QCA behavior, it is not necessary to achieve great accuracy in predicting bond lengths. QCA employs a fairly simple feature of the molecular system: is the charge on this end or on that end? Thus simpler techniques (like UHF) can do rather well.

An extension of the Aviram molecule to three dots provides a first example of clocking in molecular QCA. A third allyl dot can be added to the Aviram molecule which acts as a null state (Fig. 11.5). Calculations confirm that a perpendicular field can switch the molecule between the active and null state (Fig. 11.6). It is a feature of molecular design as to whether the molecule in zero field is in the null state ("normally null") or in an active state ("normally active"). A perpendicular clocking field can be straightforwardly constructed by using metallic clocking wires in the layer

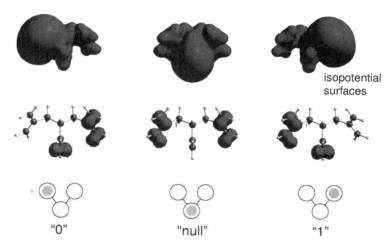

Fig. 11.5. A three-dot clocked QCA molecule. The hole occupies either the middle dot (null) or one of the two active dots (0 or 1).

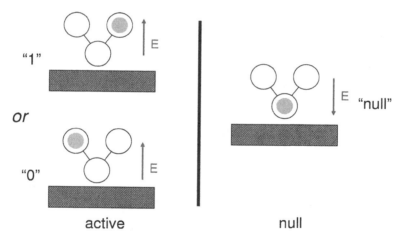

Fig. 11.6. Molecular clocking. A local perpendicular electric field switches the molecule between the active and null state.

beneath the QCA molecules (Fig. 11.7). This clocking field produces at the local molecular scale a sinusoidal wave which slides the information in the QCA layer from input to output. Complex patterning of the clocking field is possible because the clocking wires need not be at molecular size scales and can be conventionally produced. The design of the clocking field at the large scale is integrally connected to the overall computational architecture.

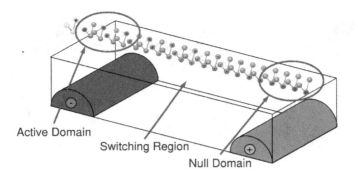

Fig. 11.7. Molecular clocking field. The potential from metallic wires beneath the molecular plane generates the clocking field which switches the molecules between the active and null states. The potential on each wire is time-varying.

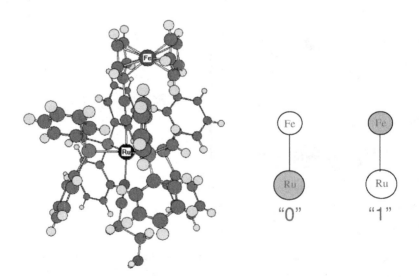

Fig. 11.8. A two-dot QCA molecule. The two unequal dots are formed by the iron center and the ruthenium center.

Fig. 11.9.   Calculated HOMO for the two-dot QCA molecule.

11.3. A Functioning Two-Dot Molecular QCA Cell

The group of Notre Dame chemist Thomas Fehlner was the first to produce a functioning two-dot QCA molecule shown in Fig. 11.8. They synthesized a mixed-valence compound with one dot formed by a ruthenium and the second dot formed by a ferrocene group.[2–4] The result is a biased double-dot — in zero field the electron is on the ruthenium dot (Fig. 11.9). A field must be applied to balance the occupation energy of the two dots. When the applied electric field is just enough to make the two energy states degenerate, the mobile electron can hop back and forth between the two dots. The associated signature is a peak in the capacitance of a layer of such double-dots when the field reaches this critical point. Measurements confirm that indeed this capacitance peak is observed, verifying the bistable switching behavior of the molecule. In fact, two peaks are observed (Fig. 11.10), corresponding to the two configurations of the neutralizing counterions. If one counterion is washed off the surface, one peak remains. This sensitivity to the local electrostatic environment is promising, though molecule-molecule interaction remains to be demonstrated.

Two key issues beyond electronic bistability have been addressed with these molecules. They have been functionalized so that they bond covalently to a silicon surface. Additionally, surrounding groups form "struts" which keep the double dot ordered in an upright configuration. These steps illustrate how careful functionalization can solve the problems of surface attachment and ordering.

Quantum chemistry calculations support the interpretation of the experiment and can go further in demonstrating molecule-molecule interaction. The calculations show field-induced switching of the dipole moment at a critical value that equalizes the double-dot energies (Fig. 11.11). In the computer, but not yet in the laboratory, it is possible to put one molecule

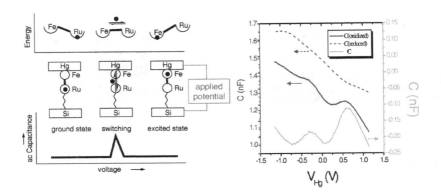

Fig. 11.10. Capacitance signature of two-dot switching. When the applied field just equalizes the energy of the two dots a peak in the capacitance is observed. The data on the right shows two peaks corresponding to the two counterion configurations present in the surface-bound species.

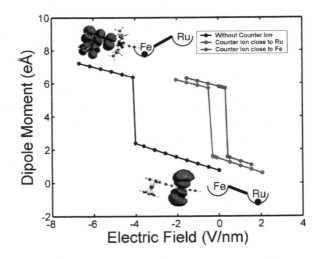

Fig. 11.11. Calculated switching of two-dot molecule by an applied electric field.

close to another and examine the interaction between them. The response of the molecular dipole moment to that of a neighboring molecule displays the desired nonlinear characteristic, as shown in Fig. 11.12.

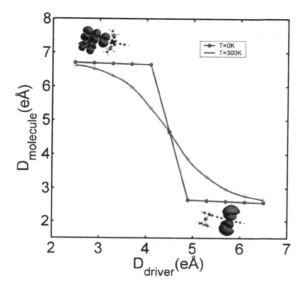

Fig. 11.12. The calculated dipole moment of a two-dot molecule as a function of the dipole moment of a neighboring molecule. The distance between the two molecules is 1nm and a 1.2 V/nm field is applied.

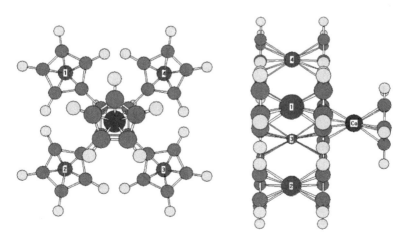

Fig. 11.13. Two views of the 4-dot QCA molecule. Atoms labeled 1-4 are the four iron centers.

"0"   "1"

Fig. 11.14.   The HOMO for the two stable states of the 4-dot QCA molecule.

11.4. A Four-Dot Molecular QCA Cell

The Fehlner group has also created a four-dot molecular QCA cell in which the four dots are each formed by a ferrocene group,[5] as shown in Fig. 11.13. The four iron atoms are held in a square by a linking complex which includes a cobalt group. Though this molecule as yet lacks functional groups for surface binding and orientation, it does represent another step in self-assembly. Chemical characterization supports the idea that a stable complex can be isolated with two mobile charges on antipodal sites of the square. Quantum chemistry calculations shown in Fig. 11.14 support this notion.

The molecule-molecule response is shown in Fig. 11.15 and confirms that the Coulomb interaction is sufficient to couple the state of one cell to a neighboring cell with the desired nonlinear response. Figures 11.16 and 11.17 show the operation, again still in calculation, of a molecular majority gate formed from these molecules.

11.5. Conclusions

It is expected that Molecular QCA will work at room temperature. We have explored theoretically model systems based on the Aviram molecule. In addition, recent work by the Fehlner group has resulted in actual molecules which demonstrate QCA switching. Theoretical extensions of the experiments done so far are extremely encouraging.

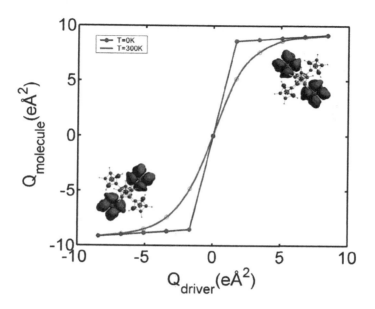

Fig. 11.15. Calculated response of the polarization (quadrupole moment) of one molecule to another molecule.

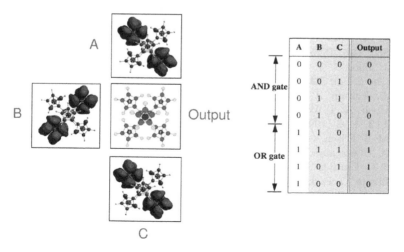

Fig. 11.16. Schematic of majority gate formed from 4-dot ferrocene molecule.

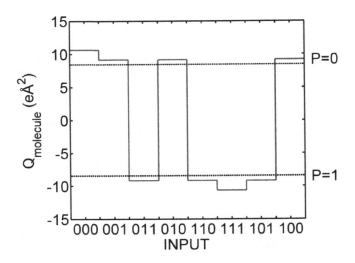

Fig. 11.17.   Calculated response of molecular majority gate shown in Fig. 11.16.

References

1. C. S. Lent, *Science* **288**, 1597 (2000).
2. X. Lei, E. E. Wolf, and T. P. Fehlner, *Eur. J. Inorg. Chem.* **1998**, 1835 (1998); W. Cen, P. Lindenfeld, T. P. Fehlner, J. Am. Chem. Soc. **114**, 5451 (1992).
3. Z. Li, A. M. Beatty, and T. P. Fehlner, **Inorg. Chem. 42**, 5707 (2003).
4. Z. Li, and T. P. Fehlner, *Inorg. Chem.* **42**, 5715 (2003); H. Qi, S. Sharma, Z. Li, G. L. Snider, A. O. Orlov, C. S. Lent, T. P. Fehlner, J. Am. Chem. Soc, **125**, 15250 (2003);
5. J. Jiao, G. J. Long, F. Grandjean, A. M. Beatty, T. P. Fehlner, *J. Am. Chem. Soc.* **125**, 7522 (2003).
6. A. Aviram, *J. Am. Chem. Soc.* **110** 5687 (1988); also discussed by N. S. Hush, A. T. Wong, G. B. Bacskay, and J. R. Reimers, *J. Am. Chem. Soc.* **112**, 4192 (1990).
7. M. J. Frisch, G. W. Trucks, H. B. Schlegel, G. E. Scuseria, M. A. Robb, J. R. Cheeseman, V. G. Zakrzewski, J. A. Montgomery, Jr., R. E. Stratmann, J. C. Burant, S. Dapprich, J. M. Millam, A. D. Daniels, K. N. Kudin, M. C. Strain, O. Farkas, J. Tomasi, V. Barone, M. Cossi, R. Cammi, B. Mennucci, C. Pomelli, C. Adamo, S. Clifford, J. Ochterski, G. A. Petersson, P. Y. Ayala, Q. Cui, K. Morokuma, P. Salvador, J. J. Dannenberg, D. K. Malick, A. D. Rabuck, K. Raghavachari, J. B. Foresman, J. Cioslowski, J. V. Ortiz, A. G. Baboul, B. B. Stefanov, G. Liu, A. Liashenko, P. Piskorz, I. Komaromi, R. Gomperts, R. L. Martin, D. J. Fox, T. Keith, M. A. Al-Laham, C. Y. Peng, A. Nanayakkara, M. Challacombe, P. M. W. Gill, B. Johnson, W. Chen, M.

W. Wong, J. L. Andres, C. Gonzalez, M. Head-Gordon, E. S. Replogle, and J. A. Pople, *Gaussian 98, Revision A.11*, Gaussian, Inc., Pittsburgh PA, 2001. Calculation done with same level of theory as Ref. [3], UHF/STO-3G.

CHAPTER 12

Magnetic Quantum-Dot Cellular Automata (MQCA)

Alexandra Imre, György Csaba[a], Gary H. Bernstein and Wolfgang Porod

Department of Electrical Engineering
University of Notre Dame
Notre Dame, IN 46556, U.S.A

12.1. Introduction

Today's computers process information electronically. Digital computation however can be done in fundamentally different physical systems as well. A method akin to quantum-dot cellular automata, based on magnetism was proposed in Ref. 1, where networks of interacting submicrometer magnetic dots are used to perform logic operations and propagate information at room temperature. Logic states are defined by the direction of magnetization of the elongated, single-domain magnetic dots. The dots couple to their nearest neighbors through magnetostatic interaction, which result in an antiferromagnetic ordering in the array of the magnets. Information flow occurs in the network by means of an externally applied oscillating magnetic field that feeds energy into the system and also serves as a clock. These networks offer a several thousandfold increase in integration density and a hundredfold reduction in power dissipation over current microelectronic technology; however, the computation speed is limited to a few hundred MHz. This chapter introduces the basic idea of computing with elongated dipole-coupled nanomagnets. This magnetic QCA concept was explored by György Csaba and Wolfgang Porod.[2,3]

The magnetic quantum-dot cellular automata concept is a version of the field-coupled QCA architecture that was first proposed in Ref. 4. The

---

[a]Current address: Institut für Nanoelektronik, Technische Universität Muenchen, Arcistrasse 21, Muenchen, Germany

original idea was introduced as a quantum dot system, in which electrons tunnel between the quantum dots affected by repelling Coulomb forces. The basic QCA geometries and their functions are shown in Fig. 12.1. Each square is an elementary building block that contains four quantum dots, one dot in each corner. With two electrons in the four-dot box, two stable states are defined by the configuration of the electrons as they occupy the dotted corners of the squares along the two diagonals. Electron tunneling occurs only inside the boxes, and the neighboring boxes are coupled by long-range electrostatic fields. This architecture can propagate and process binary information when a suitable clocking method is applied.[5]

From a more general view, the field-coupled QCA architecture is a signal processing system built from simple, identical, bistable units that are locally connected to each other solely by electromagnetic forces. Consequently, a particular signal processing function is defined by the specific placement of the building blocks.

The QCA concept can be realized in different physical systems. The first experimental demonstration of a working electronic QCA cell was published in 1997,[6] a logic gate in 1999,[7] and a shift register in 2003,[8] as described in more detail elsewhere in this volume. In this approach, the quantum dots are realized by aluminum islands separated by a few-nanometer thick

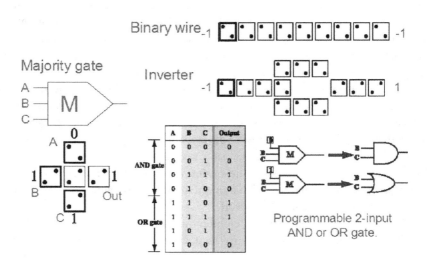

Fig. 12.1.   Quantum-dot Cellular Automata devices.

layer of aluminum-oxide. Good agreement with the theoretical results was demonstrated at 50 mK. Room temperature operation is expected only when the sizes are reduced to the molecular scale,[9] however the realization of molecular QCA requires more advanced fabrication technology.

Another room temperature approach uses magnetic dot-arrays. It has already been demonstrated that propagating magnetic excitations (domain walls, solitons) in magnetic nanostructures are suitable for digital signal processing.[11] As our recent work suggests, nanomagnets with a sufficiently large anisotropy, low coercivity, and a remanence that equals their saturation state may be utilized in a practical application as stray-field-coupled QCA building blocks. Advantages of this approach are the possibility to achieve high integration densities and low power operation/switching capabilities.

12.2. Magnetic QCA Structures

The magnetic QCA concept utilizes magnetic dipole interactions, which are caused by the stray field of submicron size magnets that are placed in an array with small separation. The digital information is represented by the magnetization states of the individual nanomagnets, just as in magnetic random access memory (MRAM) devices. To achieve this, the nanomagnets are required to be magnetically bistable, as shown in Fig. 12.2(a), so that the bit values '0' and '1' can be assigned to the stable states. This can be achieved by using uniaxial magnetic materials or suitably-shaped magnetic building blocks with a sufficiently high shape anisotropy.

Consider an elongated-shaped magnet with a hysteresis curve sketched above. Placing many of these magnets side by side along the longer axis results in an array of dipole-coupled nanomagnets that favor antiparallel alignment of their individual magnetization directions. This is due to the external field of the magnets acting on each other in close vicinity, thus tending to magnetize in the opposite direction. The antiparallel alignment that develops is called antiferromagnetic ordering, and is used during clocking of the array. Figure 12.2(b) illustrates a nanomagnet array in which the information propagates from left to right under the influence of input and output devices. The input and output of the field-coupled computing array can be realized in a similar fashion to an MRAM, but note that the magnetic dots at the interior of the array are influenced only by the fields of their neighbors, and are not accessed externally.

The field-coupled QCA devices are operated by an external clock, which

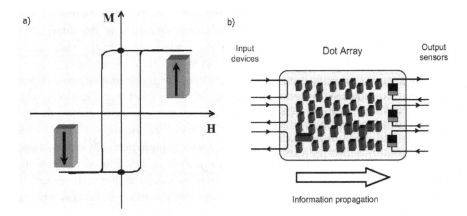

Fig. 12.2.   Schematic hysteresis curve of an elongated nanomagnet (a), and illustration of a magnetic computing array (b). Information flows from the input devices toward the output devices via magnetic interactions.

in this case is a periodically oscillating external magnetic field that drives the system from an initial state into the ground state. For a simple two-magnet system, the role of the clock is to overcome a single energy barrier, as shown in Fig. 12.3(a). The figure shows the total magnetostatic energy of two coupled single-domain permalloy nanomagnets as a function of the magnetization of the right dot. The magnetization of the left magnet is kept fixed. The magnetostatic energy is dominated by shape anisotropy of the elongated magnets, which keeps the magnetization of the dots bistable and results in two deep energy minima for magnetization directions parallel to the magnetic easy axis (for 90° and for 270°). The ground state is found for 270° due to the dipolar coupling, which favors the antiparallel alignment; while the parallel alignment is found for 90° and defines a metastable state. The energy difference between ground state and metastable state is approximately $150kT$ at room temperature, and the energy barrier from metastable to ground state is approximately $100kT$. Since both are significantly large, this system is stable at room temperature in both the metastable and ground states.

Clocking can be performed by applying a magnetic field along the magnetic hard axis of the dots, as schematically shown in the right panel of Fig. 12.3. In part (b), the external magnetic field turns the magnetic moments of all particles horizontally into a "neutral" logic state. This is an unstable state of the system, and as the field is removed, the nanomagnets

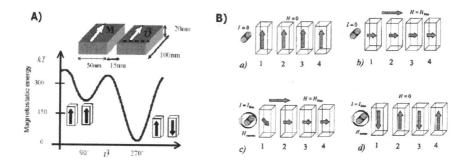

Fig. 12.3. Room temperature energy diagram of two coupled nanomagnets (A), and operating scheme of the nanowire (B): Initial configuration (a), high-field state before (b), after the application of the input (c), and (d) final ordered state.

relax into the antiferromagnetically-ordered ground state (d). If the first dot of the chain is influenced by an input device during this relaxation process, then it induces switching that sets the state of the whole chain. One can think of this as an inverter chain. Chains containing an even number of dots simply transmit the state of the input, and those containing an odd number of dots operate as an inverter. The clock affects (and evaluates) the whole chain simultaneously. The critical issue for practical applications of magnetic QCA is the length over which the antiferromagnetic ordering remains perfect (errorless). This critical length must be sufficiently large to support complicated arrangements of dots that work as a logic system.

The majority gate is the universal logic gate for field-coupled QCA devices. A top view of the layout of a magnetic majority gate built from pillars is sketched in Fig. 12.4(a). Shown here, the inputs are determined by external driving currents. In the ground state, dot '1', the central dot, will be antiparallel to the majority of its neighbors. If one of the input dots is held in the logical '1' state, the central dot performs a logical NOR function between the other two inputs (and the output). If one input is set to the logical '0' state, then the gate computes a NAND function. The truth-table of the majority gate is shown in Fig. 12.1.

12.3. Modeling of Magnetic QCA Arrays

Modeling of an interacting nanomagnet array is discussed in Refs. [2,12]. In general, the full micromagnetic equations must be solved numerically, e.g. by the standard OOMMF code.[13] For dot sizes below approximately 100

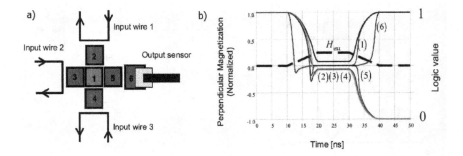

Fig. 12.4.   Layout of the magnetic majority gate. The long axis of the pillar-shaped dots is perpendicular to the plane of the paper (a). Single-domain SPICE simulation of a magnetic majority gate (b). The numbers denote the waveforms of the corresponding dot in (a), the z component of magnetization is shown, and the dashed line represents the external pumping field.

nm, the single-domain approximation (SDA) can be utilized in simulations, which results in a significant simplification. Within the framework of the SDA, the magnetization of the dots is represented by single vectors, rather than vector fields. This is applicable because magnetic particles of this size exhibit single magnetic domain behavior; the domain wall is simply the physical border, and a change in magnetization evolves through coherent rotation of atomic magnetic moments inside the particle. The SDA leads to a system of coupled ordinary differential equations, which formally resemble the dynamical equations describing the behavior of electrical circuits. Using this analogy, one can use a standard circuit simulator, such as SPICE, to model the evolution of "magnetic signals" in nanomagnet circuits. The SDA provides an intuitive, computationally efficient method for understanding and designing coupled nanomagnet arrays.

Figure 12.4(b) shows an example of a time-dependent SPICE simulation of the majority gate for a particular combination of inputs. Note that the geometry parameters of dot "6" are different from those of the other dots in the structure. This assures that this dot is the first to make the transition to the neutral logic state, and the last to switch back to a definite logic value. This fact determines the direction of signal propagation and defines dot "6" as the output of the structure.

In addition, the energy flow between adjacent dots can be understood using the SDA. HSPICE simulations indicate that each nanomagnet amplifies the power flowing along the signal path.[2] The energy of the current pulse that a magnet passes to its neighbor toward the output is significantly larger than that of the pulse received from its neighbor toward the input. In effect, each nanomagnet directs energy from the external pump field to the signal path, so that the pump field serves both as an energy source and as a system clock.

12.4. Conclusion

Technologies for realizing arrays of nanomagnets are currently under development for ultra-high density hard disk drives. Our proposed devices can be built on this existing technological base. However, the geometry of the array must be carefully designed in order to exhibit the desired functionality.

In summary, the concept of magnetic quantum-dot cellular automata implements arrays of field-coupled nanomagnets whose magnetization can be switched between two stable states. The nanomagnets are placed with their magnetic easy axis parallel to each other in the arrays, so that the ground state of the interacting system is antiferromagnetically ordered. An external time-varying magnetic field is applied to control the magnetic relaxation of the system to the ground state. The relaxation itself is used to propagate and process information. The coupled nanomagnets are capable of directing energy from the pump field to a magnetic signal path, so they are locally-active components. Logical functionality can be achieved by certain physical arrangements of the nanomagnets. The inverter and the majority gate together are sufficient to realize any arbitrary logic function between inputs and outputs.

References

1. György Csaba, Alexandra Imre, Gary H. Bernstein, Wolfgang Porod, and Vitali Metlushko, *IEEE Trans. on Nanotechnology* **1**, 209 (2002).
2. György Csaba, *Computing with field-coupled nanomagnets*, Ph.D. dissertation, University of Notre Dame, 2004.
3. G. Csaba and W. Porod, *Journal of Computational Electronics* **1**, 87 (2002).
4. C. S. Lent, P. D.Tougaw, W. Porod and G. H. Bernstein, *Nanotechnology* **4**, 49 (1993).
5. A. I. Csurgay, W. Porod, and C. S. Lent, *IEEE Trans. On Circuits and Systems I* **47**, 1212 (2000).
6. A. O. Orlov, I. Amlani, G. H. Bernstein, C. S. Lent, and G. L. Snider, *Science* **277**, 928 (1997).
7. I. Amlani, A. O. Orlov, G. Toth, G. H. Bernstein, C. S. Lent, and G. L. Snider, *Science* **284**, 289 (1999).
8. R. K. Kummamuru, A. O. Orlov, R. Ramasubramaniam, C.S. Lent, G. H.

Bernstein, and G. L. Snider, *IEEE Trans. On Electron Devices* **50**, 1906 (2003).

9. C. S. Lent and B. Isaksen, *IEEE Trans. On Electron Devices* **50**, 1890 (2003).

10. R. P. Cowburn and M. E. Welland, *Science* **287**, 1466 (2000).

11. D. A. Allwood, Gang Xiong, M. D. Cooke, C. C. Faulkner, D. Atkinson, N. Vernier and R. P. Cowburn, *Science* **296**, 2003 (2002).

12. G. Csaba, W. Porod, and A. I. Csurgay, *International Journ. Circuit Theory and Applications* **31**, 67 (2003).

13. M. J. Donahue and D. G. Porter, *OOMMF User's Guide, Version 1.0, Interagency Report NISTIR 6376* http://math.nist.gov/oommf/

CHAPTER 13

Final Remarks and Future Perspectives

Massimo Macucci

Dipartimento di Ingegneria dell'Informazione
Università di Pisa
Via Caruso 16, I-56122 Pisa, Italy

In the preceding chapters, we have provided an overview of the work performed within the QUADRANT project and within some other initiatives aimed at the evaluation of the potential of QCA circuits for large scale applications, and, possibly, at finding a route to the practical implementation of such circuits.

Several different approaches have been taken into consideration: implementations based on the GaAs/AlGaAs material system, on Silicon-On-Insulator, on metal islands connected by tunneling barriers, on nanomagnets and on molecules. From detailed modeling and from the experimental results, we have observed that semiconductor or metal implementations can be used to demonstrate that the basic principle of QCA switching is sound, but they are affected by two main limitations: the operating temperature and the need for careful tuning of each cell. Both of these limitations derive from the very small value of the electrostatic energy splitting, i.e. of the difference between the energies corresponding to the ground state cell configuration and to the first excited state. Such a splitting is, for cells of about 100 nm, of the order of a few tens of millielectronvolts, which, in the best hypothesis, limits the operating temperature to a few tens of kelvins. For the current metal island implementation, the energy splitting is reduced by an order of magnitude or more, and the same happens for the operating temperature (at present below 1 K). This also means that stray charges, geometrical imperfections, undesired electrostatic couplings (such as between biasing leads and unrelated dots) may easily disrupt the operation of a practical QCA cell.

The electrostatic energy splitting can be increased by reducing the size of the cell (it scales linearly with the inverse of cell size), but such a reduction, for more "traditional" (semiconductor and metal based) implementations, is associated with an increase of the relative fabrication errors and of the technological challenges. For semiconductor cells, in particular, the confinement energy in the dots scales with the square of the inverse of the dot size, thereby increasing faster than the electrostatic interaction. This results in a larger confinement energy shift for the same relative fabrication error, with the consequence of making adjustment requirements even stricter as cell size is shrunk down.

Molecular implementations do not suffer from fabrication tolerances, in the sense that molecules of a given type are all identical, as a result of chemical properties, but imperfections in the substrate on which they will be attached or in the host material in which they will be embedded may lead to severe asymmetry problems. Molecules proposed for the implementation of QCA circuits have relatively small dimensions, therefore benefit from an increase in the value of the electrostatic energy splitting, however the electrostatic interaction between neighbors undergoes screening by the electrons in the orbitals surrounding the redox centers that act as dots, which reduces intermolecular coupling. Some encouraging experimental results have been obtained, showing that an external electric field can switch the polarization state of a "double-dot" molecule. Further research work is needed to fully evaluate the technological potential of molecular QCA architectures.

Nanomagnetic implementations may have a potential because of the strength of the magnetic dipole interaction between neighboring nanomagnets, which makes the implementation of functional QCA circuits with cell size in the 100 nm range operating at room temperature conceivable. Although clock speeds beyond a few hundred megahertz cannot probably be achieved, nanomagnetic QCA circuits could find an application in situations requiring limited computing power and extremely low power consumption, as long as efficient ways to generate the clocking field are found.

For all implementations of QCA logic, except for the very simplest, some sort of clocking scheme appears to be necessary, if reliable operation is to be achieved: this adds quite significant complexity to the design and, in particular, to the implementation of logic circuits, since a multiphase clock needs to be distributed throughout the circuit. In principle, the best result would be obtained by reaching with a clock signal each individual cell, which would require very complex wiring, thereby making one of the main advantages of the QCA architecture, its "wireless" nature, disappear.

An intermediate solution can be achieved by assigning each clock phase to an entire region, containing a number of cells rather than to a single cell: this allows reducing the number of interconnects and simplifying the overall layout, but a price in terms of performance degradation must be paid, because within each clock zone problems similar to those typical of non-clocked QCA circuits will be present. Nevertheless, this appears to be the only viable approach, for example, in the case of molecular QCA architectures: reaching each single molecule with an electrode (a carbon nanotube has been sometimes suggested as a possible local control electrode for cells at the molecular level) would forfeit much of the simplicity that makes QCA circuits attractive.

An important lesson we have learned from the research activity within QUADRANT is that evaluating the potential of a new proposed technology is not a simple task: the total effort can be reduced if realistic modeling support is available at an early stage. Modeling tools with good quantitative predictive capabilities are instrumental in determining the expected behavior of a device well in advance of its practical implementation. On the other hand, basic models, capturing only the physical principles of operation, are often insufficient to detect and to quantify many of the problems associated with the much more complex structure of real devices. In the QUADRANT work, detailed models have shown how critical the role of fabrication tolerances is in a semiconductor implementation of the QCA paradigm and how far current technological capabilities are from satisfying the minimal technological requirements.

The formidable difficulties involved in the practical application of QCA technology should not diminish its importance as the first well thought out and conceptually sound departure from the three-terminal device paradigm that has so far dominated information processing. The current lack of device concepts and proposals likely to take up the central role of CMOS transistors beyond the "red brick wall" predicted by the ITRS roadmap hints at the need for a change of paradigm, for disruptive innovations capable of opening up new perspectives for treating information at the nanoscale. We do not know yet what the successor of CMOS technology will be, neither do we know whether it will be as successful, however we believe that in the quest for it the lessons learned from the QUADRANT work will turn out useful both for the developed methodology, in terms of close interaction between experimental activities and realistic modeling, and for the technical results, which are applicable to a wide range of charge-based nanoelectronic devices.

INDEX